JOURNAL OF APPLIED LOGICS - IFCOLOG JOURNAL OF LOGICS AND THEIR APPLICATIONS

Volume 8, Number 8

September 2021

Disclaimer

Statements of fact and opinion in the articles in Journal of Applied Logics - IfCoLog Journal of Logics and their Applications (JALs-FLAP) are those of the respective authors and contributors and not of the JALs-FLAP. Neither College Publications nor the JALs-FLAP make any representation, express or implied, in respect of the accuracy of the material in this journal and cannot accept any legal responsibility or liability for any errors or omissions that may be made. The reader should make his/her own evaluation as to the appropriateness or otherwise of any experimental technique described.

ISBN 978-1-84890-377-7
ISSN (E) 2631-9829
ISSN (P) 2631-9810

College Publications
Scientific Director: Dov Gabbay
Managing Director: Jane Spurr

http://www.collegepublications.co.uk

EDITORIAL BOARD

SCOPE AND SUBMISSIONS

This journal considers submission in all areas of pure and applied logic, including:

pure logical systems
proof theory
constructive logic
categorical logic
modal and temporal logic
model theory
recursion theory
type theory
nominal theory
nonclassical logics
nonmonotonic logic
numerical and uncertainty reasoning
logic and AI
foundations of logic programming
belief change/revision
systems of knowledge and belief
logics and semantics of programming
specification and verification
agent theory
databases

dynamic logic
quantum logic
algebraic logic
logic and cognition
probabilistic logic
logic and networks
neuro-logical systems
complexity
argumentation theory
logic and computation
logic and language
logic engineering
knowledge-based systems
automated reasoning
knowledge representation
logic in hardware and VLSI
natural language
concurrent computation
planning

This journal will also consider papers on the application of logic in other subject areas: philosophy, cognitive science, physics etc. provided they have some formal content.

Submissions should be sent to Jane Spurr (jane@janespurr.net) as a pdf file, preferably compiled in LaTeX using the IFCoLog class file.

CONTENTS

ARTICLES

PREFACE TO INTUITIONISTIC MODAL LOGIC 2017

VALERIA DE PAIVA

SERGEI ARTEMOV

The series of workshops on Intuitionistic Modal Logic and Applications (IMLA) exists because we hope that philosophers, mathematical logicians and computer scientists will share information and tools when investigating intuitionistic modal logics and modal type theories. The workshop has fulfilled at least some of its main mission, since the tools of modal logic are now part of the arsenal of Programming Language researchers, while modal logicians have become used to the possibilities of implementing their modal logic designs and models. Despite that, there are still many logicians and computer scientists who have never heard of intuitionistic modal logics and for those the collection of reports of previous IMLA meetings, taken together can be very useful.

Four years have gone since the last Intuitionistic Modal Logic and Applications workshop in July 2017 in Toulouse, France, as part of the European Summer School on Logic, Language and Information (ESSLLI). These years have been marked by upheavals, both at the personal level to the world level, given the recent devastation caused the coronavirus crisis everywhere. Hence progress on our work in constructive modal logic and applications has been slow.

There are still big questions in the field that hold broad interest and that have not been answered. Here, following Stewart, de Paiva, and Alechina [4] we understand Intuitionistic Modal Logic and Applications broadly: not only modal theories that are intuitionistic, but also constructive proof-theoretic or type-theoretic semantics, for all kinds of modal-like systems. There are many systems of constructive modal logics, even more so than the ones for classical modal logic. This much is expected, as the constructivization of a notion, usually creates several alternative possibilities that need to be compared. Some relationships are known between systems, e.g. the famous "cube" of modal logics, appearing in textbooks and the encyclopedia [1] has also been considered, recently, for intuitionistic and constructive modal logics [3]. But it seems fair to say that we do not have as a clear picture of the field, as one has for classical modal logic. Also some strands of work only make sense in the constructive setting.

The workshop had a surprising spread of work, on different approaches to modal logic. This special issue discusses five representatives of these various strands of work. First, the work of Balbiani, Boudou, Dieguez and Fernandez-Duque in "Bisimulations for intuitionistic temporal logics" speaks to Prior's tradition of considering temporal logics as akin to modal logics. Here they discuss temporal logics operators 'next' [2], 'until' and 'release' over a constructive basis and show that some of the traditional definitions of 'eventually' and 'henceforth' in terms of other operators do not hold in this constructive setting.

Secondly, Kavvos work on "Intensionality, Intensional Recursion, and the Gödel-Löb axiom" sits with Pfenning's school of constructive modal logic, based on the design of expressive type systems for practical programming languages. This aims at capturing more program errors at compile-time without sacrificing conciseness or efficiency of programs. Kavvos investigates the system obtained by adding to the Gödel-Löb axiom a novel constructive reading: it affords the programmer the possibility of a very strong kind of recursion which enables them to write programs that have access to their own code.

Thirdly, Kuznets, Marin, and Strassburger's work on "Justification logic for constructive modal logic" sits squarely within the justification logic framework. Artemov's Justification Logics (including the original Logic of Proofs) is a program still unfolding since the early 90s. Justification logic is a refinement of modal logic which studies the concepts of knowledge, belief, and provability. However, instead of simple modalities, we have a family of justification terms, associated to the modal notions. Thus, while a modal formula $\Box A$ can be read as A is known/believed, or A is provable, a justification counterpart $t : A$ of this formula is read as A is known/believed for reason t or t is evidence for A, where t is a justification term. By introducing operations on justification terms, justification logic studies the operational contents of modalities in various modal logics. However, these terms were originally conceived for necessity-like operators. The paper here extends justification terms to possibility modalities, adding in witness terms. These possibility modalities, like most in constructive modal logics, are not strictly dual to the necessity ones.

Fourthly, the work in the algebraic tradition of constructive modal logic is carried on strongly by Shapirovsky's revisiting of Glivenko's theorem for modal logics. The paper "Glivenko's theorem, finite height, and local tabularity" uses Glivenko's famous theorem (a formula is derivable in classical propositional logic iff its double negation translation is derivable in intuitionistic propositional logic) conceived in terms of heights of Kripke frames to produce a modal generalization of Glivenko's theorem.

Finally the work on "A Modal Characterisation of an Intuitionistic I/O Operation" by Parent brings to the fore a different strand of classical modal logic, the one based on deontic logic of a specific tradition, the input/output (I/O) logics. The input/output logics of Makinson and van der Torre explain the principles of deontic logic, not by some set of possible worlds among which some are ideal or at least better than others, but by reference to a set of explicit norms or existing standards. This has proved very useful when discussing issues such as the question of how to model permissions, how to accommodate and resolve conflicts between norms, and the question of how to reason about norm violation. Input/Output logic is discussed mostly for classical logic and its operations can be reformulated in terms of classical modal logic. Parent's paper provides a similar reformulation for a system of intuitionistic modal logic. The editors would like to thank the authors and reviewers for all the work they put into this special issue. We hope that the unhappy coronavirus circumstances of 2020 will improve and that the new year will bring an end to the health crisis. Hoping is, after all, allowed and free.

References

[1] James Garson. Modal logic. In Edward N. Zalta, editor, *The Stanford Encyclopedia of Philosophy*. Summer 2014 edition, 2014.

[2] Kensuke Kojima and Atsushi Igarashi. Constructive linear-time temporal logic: Proof systems and Kripke semantics. *Information and Computation*, 209(12):1491–1503, 2011.

[3] Sonia Marin and Lutz Straßburger. Label-free modular systems for classical and intuitionistic modal logics. In *Advances in Modal Logic* 10, 2014.

[4] Charles Stewart, Valeria de Paiva, and Natasha Alechina. Intuitionistic modal logic: a 15-year retrospective. *Journal of Logic and Computation*, 2015.

Received December 2020

Bisimulations for Intuitionistic Temporal Logics

Philippe Balbiani
IRIT, Toulouse University, Toulouse, France
`philippe.balbiani@irit.fr`

Joseph Boudou
IRIT, Toulouse University, Toulouse, France
`joseph.boudou@irit.fr`

Martín Diéguez
LERIA, Université d'Angers, Angers, France
`martin.dieguezlodeiro@univ-angers.fr`

David Fernández-Duque[*]
Department of Mathematics WE16, Ghent University, Ghent, Belgium
`DavidFernandezDuque@UGent.be`

Abstract

We introduce bisimulations for the logic $\mathsf{ITL}^{\mathrm{e}}$ with $\bigcirc$ ('next'), $\mathcal{U}$ ('until') and $\mathcal{R}$ ('release'), an intuitionistic temporal logic based on structures $(W, \preccurlyeq, S)$, where $\preccurlyeq$ is used to interpret intuitionistic implication and S is a $\preccurlyeq$-monotone function used to interpret the temporal modalities. Our main results are that $\Diamond$ ('eventually'), which is definable in terms of $\mathcal{U}$, cannot be defined in terms of $\bigcirc$ and $\square$, and similarly that $\square$ ('henceforth'), definable in terms of $\mathcal{R}$, cannot be defined in terms of $\bigcirc$ and $\mathcal{U}$, even over the smaller class of here-and-there models.

This research was partially supported by ANR-11-LABX-0040-CIMI within the program ANR-11-IDEX-0002-02.

[*]David Fernández-Duque's research is partially funded by the SNSF-FWO Lead Agency Grant 200021L_196176 (SNSF)/G0E2121N (FWO).

1 Introduction

The definition and study of full combinations of modal [5] and intuitionistic [6, 23] logics can be quite challenging [30], and temporal logics, such as LTL [28], are no exception. Some intuitionistic analogues of temporal logics have been proposed, including logics with 'past' and 'future' tenses [9], or with 'next' [7, 19] and 'henceforth' [17]. We proposed an alternative formulation in [4], where we defined the logics ITL^e and ITL^p using semantics similar to those of *expanding* and *persistent* products of modal logics, respectively [13], and the tenses $\circ$ ('next'), $\diamond$ ('eventually'), and $\square$ ('henceforth'). ITL^e in particular differs from previous proposals (e.g. [9, 27]) in that we consider minimal frame conditions that allow for all formulas to be upward-closed under the intuitionistic preorder, which we denote $\preccurlyeq$. We then showed that ITL^e with $\circ$ ('next'), $\diamond$ ('eventually'), and $\square$ ('henceforth') is decidable, thus obtaining the first intuitionistic analogue of LTL which contains the three tenses, is conservative over propositional intuitionistic logic, is interpreted over unbounded time, and is known to be decidable.

Note that both $\diamond$ and $\square$ are taken as primitives, in contrast with the classical case, where $\diamond\varphi$ may be defined by $\diamond\varphi \equiv \neg\square\neg\varphi$, whereas the latter equivalence is not intuitionistically valid. The same situation holds in the more expressive language with $\mathcal{U}$ ('until'): while the language with $\circ$ and $\mathcal{U}$ is equally expressive to classical monadic first-order logic with $\leq$ over $\mathbb{N}$ [12], $\mathcal{U}$ admits a first-order definable intuitionistic dual, $\mathcal{R}$ ('release'), which cannot be defined in terms of $\mathcal{U}$ using the classical definition. However, this is not enough to conclude that $\mathcal{R}$ cannot be defined in a different way. Thus, while in [4] we explored the question of decidability, here we will focus on *definability;* which of the modal operators can be defined in terms of the others?

Following Simpson [30] and other authors, we interpret the language of ITL^e using bi-relational structures, with a partial order $\preccurlyeq$ to interpret intuitionistic implication, and a function or relation, which we denote S, representing the passage of time. Alternatively, one may consider topological interpretations [8], but we will not discuss those here. Various intuitionistic temporal logics have been considered, using variants of these semantics and different formal languages. The main contributions include:

- Davies' intuitionistic temporal logic with $\circ$ [7] was provided Kripke semantics and a complete deductive system by Kojima and Igarashi [19].

- Logics with $\circ, \square$ were axiomatized by Kamide and Wansing [17], where $\square$ was interpreted over bounded time.

- Nishimura [25] provided a sound and complete axiomatization for an intuitionistic variant of the propositional dynamic logic PDL.

- Balbiani and Diéguez axiomatized the here-and-there variant of LTL with $\bigcirc, \Diamond, \square$ [2], here denoted $\mathsf{ITL^{ht}}$.

- Fernández-Duque [10] proved the decidability of a logic based on topological semantics with $\bigcirc, \Diamond$ and a universal modality.

- The authors [4] proved that the logic $\mathsf{ITL^e}$ with $\bigcirc, \Diamond, \square$ has the strong finite model property and hence is decidable, yet the logic $\mathsf{ITL^p}$, based on a more restrictive class of frames, does not enjoy the fmp.

In this paper, we extend $\mathsf{ITL^e}$ to include $\mathcal{U}$ ('until') and $\mathcal{R}$ ('release'). As is well-known, $\Diamond\varphi \equiv \top\mathcal{U}\varphi$ and $\square\varphi \equiv \bot\mathcal{R}\varphi$; these equivalences remain valid in the intuitionistic setting, but many of the tenses are no longer inter-definable as in the classical case. To show this, we will introduce different notions of bisimulation which preserve formulas with $\bigcirc$ and each of $\Diamond, \square, \mathcal{U}$ and $\mathcal{R}$. With this, we will show that $\mathcal{R}$ (or even $\square$) may not be defined in terms of $\mathcal{U}$ over the class of here-and-there models, while $\Diamond$ *can* be defined in terms of $\square$, and $\mathcal{U}$ can be defined in terms of $\mathcal{R}$ over this class. However, we show that over the wider class of expanding models, $\Diamond$ cannot be defined in terms of $\square$.

2 Syntax and semantics

We will work in sublanguages of the language $\mathcal{L}$ given by the following grammar:

$$\varphi, \psi := p \mid \bot \mid \varphi \wedge \psi \mid \varphi \vee \psi \mid \varphi \to \psi \mid \bigcirc\varphi \mid \Diamond\varphi \mid \square\varphi \mid \varphi\mathcal{U}\psi \mid \varphi\mathcal{R}\psi$$

where p is an element of a countable set of propositional variables $\mathbb{P}$. All sublanguages we will consider include all Boolean operators and $\bigcirc$, hence we denote them by displaying the additional connectives as a subscript; for example, $\mathcal{L}_{\Diamond\square}$ denotes the $\mathcal{U}$-free, $\mathcal{R}$-free fragment. As an exception to this general convention, $\mathcal{L}_{\bigcirc}$ denotes the fragment without $\Diamond, \square, \mathcal{U}$ or $\mathcal{R}$. As in the propositional case, $\neg\varphi \overset{def}{=} \varphi \to \bot$.

Given any formula φ, we define the *length* of φ (in symbols, $|\varphi|$) recursively as follows:

- $|p| = |\bot| = 0$;
- $|\phi \odot \psi| = 1 + |\phi| + |\psi|$, with $\odot \in \{\vee, \wedge, \to, \mathcal{R}, \mathcal{U}\}$;
- $|\odot\psi| = 1 + |\psi|$, with $\odot \in \{\neg, \bigcirc, \square, \Diamond\}$.

Broadly speaking, the length of a formula φ corresponds to the number of connectives appearing in φ.

2.1 Dynamic posets

Formulas of $\mathcal{L}$ are interpreted over dynamic posets. A *dynamic poset* is a tuple $\mathcal{D} = (W, \leqslant, S)$, where W is a non-empty set of states, $\leqslant$ is a partial order, and S is a function from W to W satisfying the *forward confluence* condition that for all $w, v \in W$, if $w \leqslant v$ then $S(w) \leqslant S(v)$. An *intuitionistic dynamic model*, or simply *model*, is a tuple $\mathcal{M} = (W, \leqslant, S, V)$ consisting of a dynamic poset equipped with a valuation function V from W to sets of propositional variables that is $\leqslant$-monotone, in the sense that for all $w, v \in W$, if $w \leqslant v$ then $V(w) \subseteq V(v)$. In the standard way, we define $S^0(w) = w$ and, for all $k > 0$, $S^k(w) = S(S^{k-1}(w))$. Then we define the satisfaction relation $\vDash$ inductively by:

1. $\mathcal{M}, w \vDash p$ iff $p \in V(w)$;

2. $\mathcal{M}, w \nvDash \bot$;

3. $\mathcal{M}, w \vDash \varphi \wedge \psi$ iff $\mathcal{M}, w \vDash \varphi$ and $\mathcal{M}, w \vDash \psi$;

4. $\mathcal{M}, w \vDash \varphi \vee \psi$ iff $\mathcal{M}, w \vDash \varphi$ or $\mathcal{M}, w \vDash \psi$;

5. $\mathcal{M}, w \vDash \bigcirc\varphi$ iff $\mathcal{M}, S(w) \vDash \varphi$;

6. $\mathcal{M}, w \vDash \varphi \rightarrow \psi$ iff $\forall v \geqslant w$, if $\mathcal{M}, v \vDash \varphi$, then $\mathcal{M}, v \vDash \psi$;

7. $\mathcal{M}, w \vDash \Diamond\varphi$ iff there exists k s.t. $\mathcal{M}, S^k(w) \vDash \varphi$;

8. $\mathcal{M}, w \vDash \Box\varphi$ iff for all k, $\mathcal{M}, S^k(w) \vDash \varphi$;

9. $\mathcal{M}, w \vDash \varphi \mathcal{U} \psi$ iff there exists $k \geq 0$ s.t. $\mathcal{M}, S^k(w) \vDash \psi$ and $\forall i \in [0, k)$, $\mathcal{M}, S^i(w) \vDash \varphi$;

10. $\mathcal{M}, w \vDash \varphi \mathcal{R} \psi$ iff for all $k \geq 0$, either $\mathcal{M}, S^k(w) \vDash \psi$, or $\exists i \in [0, k)$ s.t. $\mathcal{M}, S^i(w) \vDash \varphi$.

As usual, a formula φ is *satisfiable over a class of models* Ω if there is a model $\mathcal{M} \in \Omega$ and a world w of $\mathcal{M}$ so that $\mathcal{M}, w \vDash \varphi$, and *valid over* Ω if, for every world w of every model $\mathcal{M} \in \Omega$, $\mathcal{M}, w \vDash \varphi$. Satisfiability (validity) over the class of models based on an arbitrary dynamic poset will be called *satisfiability (validity)* for ITL^e, or *expanding domain linear temporal logic*.[1]

The relation between dynamic posets and expanding products of modal logics is detailed in [4], where the following is also shown. Below, we use the notation $\llbracket \varphi \rrbracket = \{w \in W \mid \mathcal{M}, w \vDash \varphi\}$.

Lemma 2.1. *Let $\mathcal{D} = (W, \leqslant, S)$, where $(W, \leqslant)$ is a poset and $S : W \rightarrow W$ is any function. Then, $\mathcal{D}$ is a dynamic poset if and only if, for every valuation V on W and every formula φ, $\llbracket \varphi \rrbracket$ is $\leqslant$-monotone, i.e., if $w \in \llbracket \varphi \rrbracket$ and $v \geqslant w$, then $v \in \llbracket \varphi \rrbracket$.*

Proof. The left to right direction is proved by induction on φ. The case of $\varphi \in \mathbb{P}$ is proved by using the condition on V. The rest of the inductive steps are routine. For

---

[1]Note that in [4] we used 'ITL^e' to denote the fragment of this logic without $\mathcal{U}, \mathcal{R}$.

instance, let us consider the case of $\varphi \mathcal{U} \psi$ and suppose that $v \succcurlyeq w$ and $w \in [\![\varphi \mathcal{U} \psi]\!]$. Then, there exists $k \geq 0$ such that $\mathcal{M}, S^k(w) \vDash \psi$ and for all $0 \leq j < k$, $\mathcal{M}, S^j(w) \vDash \varphi$. Since S is confluent, an easy induction shows that $S^i(v) \succcurlyeq S^i(w)$ for all $0 \leq i \leq k$. Therefore, by induction hypothesis, we get $\mathcal{M}, S^k(v) \vDash \psi$ and for all $0 \leq j < k$, $\mathcal{M}, S^j(v) \vDash \varphi$, hence $v \in [\![\varphi \mathcal{U} \psi]\!]$. For the converse direction we assume that $\mathcal{D} = (W, \preccurlyeq, S)$ and $w, v \in W$ such that $v \succcurlyeq w$ and $S(w) \not\preccurlyeq S(v)$. Take $p \in \mathbb{P}$ and define $V(u) = \{p\}$ if $S(w) \preccurlyeq u$, $V(u) = \varnothing$ otherwise. It is easy to see that V is $\preccurlyeq$-monotone, but $p \notin V(S(v))$ (because $S(w) \not\preccurlyeq S(v)$) and $p \in V(S(w))$ (because $S(w) \preccurlyeq S(w)$), from which it follows that $(\mathcal{D}, V), w \vDash \bigcirc p$ but $(\mathcal{D}, V), v \nvDash \bigcirc p$. $\qquad \square$

This suggests that dynamic posets provide suitable semantics for intuitionistic LTL. Moreover, dynamic posets are convenient from a technical point of view:

Theorem 2.2 ([4]). *There exists a computable function B such that any formula $\varphi \in \mathcal{L}_{\Diamond \Box}$ satisfiable (resp. falsifiable) on an arbitrary model is satisfiable (resp. falsifiable) on a model whose size is bounded by $B(|\varphi|)$.*

It follows that the $\mathcal{L}_{\Diamond \Box}$-fragment of ITL^e is decidable. Moreover, as we will see below, many of the familiar axioms of classical LTL are valid over the class of dynamic posets, making them a natural choice of semantics for intuitionistic LTL.

2.2 Persistent posets

Despite the advantages of dynamic posets, in the literature one typically considers a more restrictive class of frames, as we define them below.

Definition 2.3. *Let $(W, \preccurlyeq)$ be a poset. If $S \colon W \to W$ is such that, whenever $v \succcurlyeq S(w)$, there is $u \succcurlyeq w$ such that $v = S(u)$, we say that S is* backward confluent. *If S is both forward and backward confluent, we say that it is* persistent. *A tuple $(W, \preccurlyeq, S)$ where S is persistent is a* persistent intuitionistic temporal frame, *and the set of valid formulas over the class of persistent intuitionistic temporal frames is denoted* ITL^p, *or persistent domain LTL.*

As we will see, persistent frames do have some technical advantages over arbitrary dynamic posets. Nevertheless, they have a crucial disadvantage:

Theorem 2.4 ([4]). *The logic* ITL^p *does not have the finite model property, even for formulas in $\mathcal{L}_{\Diamond \Box}$.*

2.3 Temporal here-and-there models

An even smaller class of models which, nevertheless, has many applications is that of temporal here-and-there models [2]. Some of the results we will present here apply to this class, so it will be instructive to review it. Recall that the logic of here-and-there is the maximal logic strictly between classical and intuitionistic propositional logic, given by a frame $\{0,1\}$ with $0 \leqslant 1$. The logic of here-and-there is obtained by adding to intuitionistic propositional logic the axiom $p \vee (p \to q) \vee \neg q$.

A temporal here-and-there frame is a persistent frame that is 'locally' based on this frame. We can define here-and-there models using the following construction.

Definition 2.5. *Let T be a set and $f\colon T \to T$. We define a dynamic poset* $\mathrm{HT}(T,f) = (W, \leqslant, S)$, *with* $W = T \times \{0,1\}$, $(t,i) \leqslant (s,j)$ *if and only if $t = s$ and $i \leq j$, and* $S(t,i) = (f(t),i)$.

The prototypical example is the frame $\mathrm{HT}(\mathbb{N}, f)$, where $f(n) = n + 1$. Note, however, that our definition allows for other values of T (see Figure 1). In [2], this logic is axiomatized, and it is shown that $\Box$ cannot be defined in terms of $\Diamond$, a result we will strengthen here to show that $\Box$ cannot be defined even in terms of $\mathcal{U}$. It is also claimed in [2] that $\Diamond$ is not definable in terms of $\Box$ over the class of here-and-there models, but as we will see in Proposition 6.3, this claim is incorrect.

3 Some valid and non-valid ITL^e-formulas

In this section we explore which axioms of classical **LTL** are still valid in our setting. We start by showing that the intuitionistic version of the interaction and induction axioms used in [2] remain valid in our setting. However, not all Fisher-Servi axioms [11], which are valid in the here-and-there LTL of [2], are valid in ITL^e.

Proposition 3.1. *The following formulas:*

1. $\bigcirc \bot \leftrightarrow \bot$

2. $\bigcirc(\varphi \wedge \psi) \leftrightarrow (\bigcirc\varphi \wedge \bigcirc\psi)$;

3. $\bigcirc(\varphi \vee \psi) \leftrightarrow (\bigcirc\varphi \vee \bigcirc\psi)$;

4. $\bigcirc(\varphi \to \psi) \to (\bigcirc\varphi \to \bigcirc\psi)$;

5. $\Box(\varphi \to \psi) \to (\Box\varphi \to \Box\psi)$;

6. $\Box(\varphi \to \psi) \to (\Diamond\varphi \to \Diamond\psi)$;

7. $\Diamond(\varphi \vee \psi) \to (\Diamond\varphi \vee \Diamond\psi)$;

8. $\Box\varphi \leftrightarrow \varphi \wedge \bigcirc\Box\varphi$;

9. $\varphi \vee \bigcirc\Diamond\varphi \leftrightarrow \Diamond\varphi$;

10. $\Box(\varphi \to \bigcirc\varphi) \to (\varphi \to \Box\varphi)$

11. $\Box(\bigcirc\varphi \to \varphi) \to (\Diamond\varphi \to \varphi)$.

are ITL^e-*valid.*

Proof. Let us consider (10) and (11). For (10), let $\mathcal{M} = (W, \leqslant, S)$ be any ITL^e model and $w \in W$ be such that $\mathcal{M}, w \vDash \Box(\varphi \to \bigcirc\varphi)$. Let $v \geqslant w$ be arbitrary and assume that $\mathcal{M}, v \vDash \varphi$. Then, by induction on i we obtain that $S^i(w) \leqslant S^i(v)$ for all i; since $\mathcal{M}, S^i(w) \vDash \varphi \to \bigcirc\varphi$ for all i, it follows that $\mathcal{M}, S^i(v) \vDash \varphi \to \bigcirc\varphi$ for all i as well. Hence an easy induction shows that $\mathcal{M}, S^i(v) \vDash \varphi$ for all i, which means that $\mathcal{M}, v \vDash \Box\varphi$. Since w was arbitrary, we conclude that the formula (10) is valid.

For (11), let $\mathcal{M}$ be as above and $w \in W$ be such that $\mathcal{M}, w \vDash \Box(\bigcirc\varphi \to \varphi)$. Let $v \geqslant w$ be such that $\mathcal{M}, v \vDash \Diamond\varphi$, and let n be least so that $\mathcal{M}, S^n(v) \vDash \varphi$. If $n > 0$ then from $\bigcirc\varphi \to \varphi$ we obtain $\mathcal{M}, S^{n-1}(v) \vDash \varphi$, contradicting the minimality of n. We conclude that $n = 0$, hence $\mathcal{M}, v \vDash \varphi$.

The proofs for the rest of formulas are standard. $\qquad\qquad\square$

Some of the well-known Fisher Servi axioms [11] are only valid on the class of persistent frames.

Proposition 3.2. *The formulas*

1. $(\bigcirc\varphi \to \bigcirc\psi) \to \bigcirc(\varphi \to \psi),$ $\qquad\qquad$ *2.* $(\Diamond\varphi \to \Box\psi) \to \Box(\varphi \to \psi)$

are not ITL^e-*valid. However they are* ITL^p-*valid.*

Proof. Let $\{p, q\}$ be a set of propositional variables and let us consider the ITL^e model $\mathcal{M} = (W, \leqslant, S, V)$ defined as: 1) $W = \{w, v, u\}$; 2) $S(w) = v$, $S(v) = v$ and $S(u) = u$; 3) $v \leqslant u$; 4) $V(p) = \{u\}$. Clearly, $\mathcal{M}, u \nvDash p \to q$, so $\mathcal{M}, v \nvDash p \to q$. By definition, $\mathcal{M}, w \nvDash \bigcirc(p \to q)$ and $\mathcal{M}, w \nvDash \Box(p \to q)$; however, it can easily be checked that $\mathcal{M}, w \vDash \bigcirc p \to \bigcirc q$ and $\mathcal{M}, w \vDash \Diamond p \to \Box q$, so $\mathcal{M}, w \nvDash (\bigcirc p \to \bigcirc q) \to \bigcirc(p \to q)$ and $\mathcal{M}, w \nvDash (\Diamond p \to \Box q) \to \Box(p \to q)$.

Let us check their validity over the class of persistent frames. For (1), let $\mathcal{M} = (W, \leqslant, S, V)$ be an ITL^p model and w a world of $\mathcal{M}$ such that $\mathcal{M}, w \vDash \bigcirc\varphi \to \bigcirc\psi$. Suppose that $v \geqslant S(w)$ satisfies $\mathcal{M}, v \vDash \varphi$. By backward confluence, there exists $u \geqslant w$ such that $v = S(u)$, so that $\mathcal{M}, u \vDash \bigcirc\varphi$ and thus $\mathcal{M}, u \vDash \bigcirc\psi$. But this means that $\mathcal{M}, v \vDash \psi$, and since $v \geqslant S(w)$ was arbitrary, $\mathcal{M}, S(w) \vDash \varphi \to \psi$, i.e. $\mathcal{M}, w \vDash \bigcirc(\varphi \to \psi)$.

Similarly, for (2) let us assume that $\mathcal{M} = (W, \leqslant, S, V)$ is an ITL^p model and w a world of $\mathcal{M}$ such that $\mathcal{M}, w \vDash \Diamond\varphi \to \Box\psi$. Consider arbitrary $k \in \mathbb{N}$, and suppose that $v \geqslant S^k(w)$ is such that $\mathcal{M}, v \vDash \varphi$. Then, it is readily checked that the composition of backward confluent functions is backward confluent, so that in particular S^k is backward confluent. This means that there is $u \geqslant w$ such that $S^k(u) = v$. But then, $\mathcal{M}, u \vDash \Diamond\varphi$, hence $\mathcal{M}, u \vDash \Box\psi$, and $\mathcal{M}, v \vDash \psi$. It follows that $\mathcal{M}, S^k(w) \vDash \varphi \to \psi$, and since k was arbitrary, $\mathcal{M}, w \vDash \Box(\varphi \to \psi)$. $\qquad\square$

We make a special mention of the schema $\Box(\Box\varphi \to \psi) \vee \Box(\Box\psi \to \varphi)$, which characterises the class of *weakly connected frames* [14] in classical modal logic. We say that a frame (W, R, V) is weakly connected iff it satisfies the following first-order property: for every $x, y, z \in W$, if $x\,R\,y$ and $x\,R\,z$, then either $y\,R\,z$, $y = z$, or $z\,R\,y$.

Proposition 3.3. *The axiom schema* $\Box(\Box\varphi \to \psi) \vee \Box(\Box\psi \to \varphi)$ *is not* $\mathsf{ITL}^{\mathsf{ht}}$-*valid.*

Proof. Let us consider the set of propositional variables $\{p, q\}$, $T = \{0, 1\}$, $f : T \to T$ be given by $f(x) = 1$, and let $\mathcal{M} = (W, \leqslant, S, V)$ be the here-and-there model based on $\mathrm{HT}(T, f)$ with $V(p) = \{(0, 1), (1, 1)\}$ and $V(q) = \{(1, 0), (1, 1)\}$. The reader can check that $\mathcal{M}, (0, 0) \not\models \Box p \to q$ and $\mathcal{M}, (0, 1) \not\models \Box q \to p$. Consequently, $\mathcal{M}, w \not\models \Box(\Box p \to q) \vee \Box(\Box q \to p)$. $\square$

Finally, we show that $\Diamond\varphi$ (resp. $\Box\varphi$) can be defined in terms of $\mathcal{U}$ (resp. $\mathcal{R}$) and the LTL axioms involving $\mathcal{U}$ and $\mathcal{R}$ are also valid in our setting:

Proposition 3.4. *The following formulas are* $\mathsf{ITL}^{\mathsf{e}}$-*valid:*

1. $\varphi\mathcal{U}\psi \leftrightarrow \psi \vee (\varphi \wedge \circ(\varphi\mathcal{U}\psi))$;
2. $\varphi\mathcal{R}\psi \leftrightarrow \psi \wedge (\varphi \vee \circ(\varphi\mathcal{R}\psi))$;
3. $\varphi\mathcal{U}\psi \to \Diamond\psi$;
4. $\Box\psi \to \varphi\mathcal{R}\psi$;
5. $\Diamond\varphi \leftrightarrow \top\mathcal{U}\varphi$;
6. $\Box\varphi \leftrightarrow \bot\mathcal{R}\varphi$;
7. $\circ(\varphi\mathcal{U}\psi) \leftrightarrow \circ\varphi\mathcal{U}\circ\psi$;
8. $\circ(\varphi\mathcal{R}\psi) \leftrightarrow \circ\varphi\mathcal{R}\circ\psi$.

Proof. We consider some cases below. For (1), from left to right, let us assume that $\mathcal{M}, w \models \varphi\mathcal{U}\psi$. Therefore there exists $k \geq 0$ s.t. $\mathcal{M}, S^k(w) \models \psi$ and for all j satisfying $0 \leq j < k$, $\mathcal{M}, S^j(w) \models \varphi$. If $k = 0$ then $\mathcal{M}, w \models \psi$ while, if $k > 0$ it follows that $\mathcal{M}, w \models \varphi$ and $\mathcal{M}, S(w) \models \varphi\mathcal{U}\psi$. Therefore $\mathcal{M}, w \models \psi \vee (\varphi \wedge \circ(\varphi\mathcal{U}\psi))$. From right to left, if $\mathcal{M}, w \models \psi$ then $\mathcal{M}, w \models \varphi\mathcal{U}\psi$ by definition. If $\mathcal{M}, w \models \varphi \wedge \circ(\varphi\mathcal{U}\psi)$ then $\mathcal{M}, w \models \varphi$ and $\mathcal{M}, S(w) \models \varphi\mathcal{U}\psi$ so, due to the semantics, we conclude that $\mathcal{M}, w \models \varphi\mathcal{U}\psi$. In any case, $\mathcal{M}, w \models \varphi\mathcal{U}\psi$.

For (2), we work by contrapositive. From right to left, let us assume that $\mathcal{M}, w \not\models \varphi\mathcal{R}\psi$. Therefore there exists $k \geq 0$ s.t. $\mathcal{M}, S^k(w) \not\models \psi$ and for all j satisfying $0 \leq j < k$, $\mathcal{M}, S^j(w) \not\models \varphi$. If $k = 0$ then $\mathcal{M}, w \not\models \psi$ while, if $k > 0$ it follows that $\mathcal{M}, w \not\models \varphi$ and $\mathcal{M}, S(w) \not\models \varphi\mathcal{R}\psi$. In any case, $\mathcal{M}, w \not\models \psi \wedge (\varphi \vee \circ(\varphi\mathcal{R}\psi))$. From left to right, if $\mathcal{M}, w \not\models \psi$ then $\mathcal{M}, w \not\models \varphi\mathcal{R}\psi$ by definition. If $\mathcal{M}, w \not\models \varphi \vee \circ(\varphi\mathcal{R}\psi)$ then $\mathcal{M}, w \not\models \varphi$ and $\mathcal{M}, S(w) \not\models \varphi\mathcal{U}\psi$ so, due to the semantics of $\mathcal{R}$, we conclude that $\mathcal{M}, w \not\models \varphi\mathcal{R}\psi$. In any case, $\mathcal{M}, w \not\models \varphi\mathcal{R}\psi$.

For (7), from left to right, let us assume that $\mathcal{M}, w \models \circ(\varphi\mathcal{U}\psi)$. Therefore there exists $k \geq 0$ s.t. $\mathcal{M}, S^{k+1}(w) \models \psi$ and for all j satisfying $0 \leq j < k$, $\mathcal{M}, S^{j+1}(w) \models \varphi$. It follows from $\mathcal{M}, S^{k+1}(w) \models \psi$ that $\mathcal{M}, S^k(w) \models \circ\psi$, and from $\mathcal{M}, S^{j+1}(w) \models \varphi$

that $\mathcal{M}, S^j(w) \vDash \bigcirc\varphi$ for all $j < k$. We conclude that $\mathcal{M}, w \vDash \bigcirc\varphi \mathcal{U} \bigcirc\psi$. Conversely, if $\mathcal{M}, w \vDash \bigcirc\varphi \mathcal{U} \bigcirc\psi$, then there is $k \geq 0$ so that $\mathcal{M}, S^k(w) \vDash \bigcirc\psi$ and, for all $i < k$, $\mathcal{M}, S^i(w) \vDash \bigcirc\varphi$. It follows that $\mathcal{M}, S^{k+1}(w) \vDash \psi$ and, for all $i < k$, $\mathcal{M}, S^{i+1}(w) \vDash \varphi$, witnessing that $\mathcal{M}, S(w) \vDash \varphi \mathcal{U}\psi$ and $\mathcal{M}, w \vDash \bigcirc(\varphi \mathcal{U}\psi)$.

For (8), we proceed similarly, but work by contrapositive. From right to left, let us assume that $\mathcal{M}, w \not\vDash \bigcirc(\varphi \mathcal{R}\psi)$. Therefore there exists $k \geq 0$ s.t. $\mathcal{M}, S^{k+1}(w) \not\vDash \psi$ and for all j satisfying $0 \leq j < k$, $\mathcal{M}, S^{j+1}(w) \not\vDash \varphi$. This implies that $\mathcal{M}, S^k(w) \not\vDash \bigcirc\psi$ and for all j satisfying $0 \leq j < k$, $\mathcal{M}, S^j(w) \not\vDash \bigcirc\varphi$, hence $\mathcal{M}, w \not\vDash \bigcirc(\varphi \mathcal{R}\psi)$. Similarly, if $\mathcal{M}, w \not\vDash \bigcirc(\varphi \mathcal{R}\psi)$ then any $k \geq 0$ so that $\mathcal{M}, S^k(w) \not\vDash \bigcirc\psi$ and, for all $i < k$, $\mathcal{M}, S^i(w) \not\vDash \bigcirc\varphi$ yields $\mathcal{M}, S^{k+1}(w) \not\vDash \psi$ and, for all $i < k$, $\mathcal{M}, S^{i+1}(w) \not\vDash \varphi$, witnessing that $\mathcal{M}, S(w) \not\vDash \varphi \mathcal{R}\psi$ and $\mathcal{M}, w \not\vDash \bigcirc(\varphi \mathcal{R}\psi)$.

The proof of the remaining items is routine. $\square$

As in the classical case, over the class of persistent models we can 'push down' all occurrences of $\bigcirc$ to the propositional level. Say that a formula φ is in $\bigcirc$-*normal form* if all occurrences of $\bigcirc$ are of the form $\bigcirc^i p$, with p a propositional variable.

Theorem 3.5. *Given* $\varphi \in \mathcal{L}$, *there exists* $\widetilde{\varphi}$ *in* $\bigcirc$-*normal form such that* $\varphi \leftrightarrow \widetilde{\varphi}$ *is valid over the class of persistent models.*

Proof. The claim can be proven by structural induction using the validities in Propositions 3.1, 3.2 and 3.4. $\square$

We remark that the only reason that this argument does not apply to arbitrary ITL^e models is the fact that $(\bigcirc\varphi \to \bigcirc\psi) \to \bigcirc(\varphi \to \psi)$ is not valid in general (Proposition 3.2).

4 Bounded bisimulations for $\diamond$ and $\square$

In this section we adapt the classical definition of bounded bisimulations for modal logic [3] to our case. To do so we combine the ordinary definition of bounded bisimulations with the work of [26] on bisimulations for propositional intuitionistic logic. Such work introduces extra conditions involving the partial order $\preccurlyeq$. In our setting, we combine both approaches in order to define bisimulation for a language involving $\diamond$, $\square$ and $\bigcirc$ as modal operators plus an intuitionistic $\to$. Since all languages we consider contain Booleans and $\bigcirc$, it is convenient to begin with a 'basic' notion of bisimulation for this language.

Definition 4.1. *Given* $n > 0$ *and two* ITL^e *models* $\mathcal{M}_1$ *and* $\mathcal{M}_2$, *a sequence of binary relations* $\mathcal{Z}_n \subseteq \cdots \subseteq \mathcal{Z}_0 \subseteq W_1 \times W_2$ *is said to be a* bounded $\bigcirc$-bisimulation *if for all* $(w_1, w_2) \in W_1 \times W_2$ *and for all* $0 \leq i < n$, *the following conditions are satisfied:*

2273

ATOMS. *If $w_1 \, \mathcal{Z}_i \, w_2$ then for all propositional variables p, $\mathcal{M}_1, w_1 \vDash p$ iff $\mathcal{M}_2, w_2 \vDash p$.*

FORTH →. *If $w_1 \, \mathcal{Z}_{i+1} \, w_2$ then for all $v_1 \in W_1$, if $v_1 \succcurlyeq w_1$, there exists $v_2 \in W_2$ such that $v_2 \succcurlyeq w_2$ and $v_1 \, \mathcal{Z}_i \, v_2$.*

BACK →. *If $w_1 \, \mathcal{Z}_{i+1} \, w_2$ then for all $v_2 \in W_2$ if $v_2 \succcurlyeq w_2$ then there exists $v_1 \in W_1$ such that $v_1 \succcurlyeq w_1$ and $v_1 \, \mathcal{Z}_i \, v_2$.*

FORTH ○. *if $w_1 \, \mathcal{Z}_{i+1} \, w_2$ then $S(w_1) \, \mathcal{Z}_i \, S(w_2)$.*

Note that there is not 'back' clause for ○; this is simply because S is a function, so its 'forth' and 'back' clauses are identical. Bounded ○-bisimulations are useful because they preserve the truth of relatively small $\mathcal{L}_○$-formulas.

Lemma 4.2. *Given two ITL^e models $\mathcal{M}_1$ and $\mathcal{M}_2$ and a bounded ○-bisimulation $\mathcal{Z}_n \subseteq \cdots \subseteq \mathcal{Z}_0$ between them, for all $0 \le i \le n$ and $(w_1, w_2) \in W_1 \times W_2$, if $w_1 \, \mathcal{Z}_i \, w_2$ then for all $\varphi \in \mathcal{L}_○$ satisfying $|\varphi| \le i^2$, $\mathcal{M}_1, w_1 \vDash \varphi$ iff $\mathcal{M}_2, w_2 \vDash \varphi$.*

Proof. We proceed by induction on i. Let $0 \le i \le n$ be such that for all $j < i$ the lemma holds. Let $w_1 \in W_1$ and $w_2 \in W_2$ be such that $w_1 \, \mathcal{Z}_i \, w_2$ and let us consider $\varphi \in \mathcal{L}_◇$ such that $|\varphi| \le i$. The cases where φ is an atom or of the forms $\theta \wedge \psi$, $\theta \vee \psi$ are as in the classical case and we omit them. Thus we focus on the following:

CASE $\varphi = \theta \to \psi$. We proceed by contrapositive to prove the left-to-right implication. Note that in this case we must have $i > 0$.

Assume that $\mathcal{M}_2, w_2 \nvDash \theta \to \psi$. Therefore there exists $v_2 \in W_2$ such that $v_2 \succcurlyeq w_2$, $\mathcal{M}_2, v_2 \vDash \theta$, and $\mathcal{M}_2, v_2 \nvDash \psi$. By the BACK → condition, it follows that there exists $v_1 \in W_1$ such that $v_1 \succcurlyeq w_1$ and $v_1 \, \mathcal{Z}_{i-1} \, v_2$. Since $|\theta| \le i - 1$ and $|\psi| \le i - 1$, by the induction hypothesis, it follows that $\mathcal{M}_1, v_1 \vDash \theta$ and $\mathcal{M}_1, v_1 \nvDash \psi$. Consequently, $\mathcal{M}_1, w_1 \nvDash \theta \to \psi$. The converse direction is proved in a similar way but using FORTH →.

CASE $\varphi = ○\psi$. Once again we have that $i > 0$. Assume that $\mathcal{M}_1, w_1 \vDash ○\psi$, so that $\mathcal{M}_1, S(w_1) \vDash \psi$. By FORTH ○, $S_1(w_1) \, \mathcal{Z}_{i-1} \, S_2(w_2)$. Moreover, $|\psi| \le i - 1$, so that by the induction hypothesis, $\mathcal{M}_2, S(w_2) \vDash \psi$, and $\mathcal{M}_2, w_2 \vDash ○\psi$. The right-to-left direction is analogous. $\qquad\square$

Next, we will extend the notion of a bounded ○-bisimulation to include other tenses. Let us begin with ◇.

---

[2] Although not optimal, we use the length of the formula in this lemma for the sake of simplicity. More precise measures like counting the number of modalities and implications could be equally used.

Definition 4.3. *Given $n > 0$ and two ITL^e models $\mathcal{M}_1$ and $\mathcal{M}_2$, a bounded $\Diamond$-bisimulation $\mathcal{Z}_n \subseteq \cdots \subseteq \mathcal{Z}_0 \subseteq W_1 \times W_2$ is said to be a* bounded $\Diamond$-bisimulation *if for all $(w_1, w_2) \in W_1 \times W_2$ and for all $0 \leq i < n$, if $w_1\,\mathcal{Z}_{i+1}\,w_2$, then the following conditions are satisfied:*

FORTH $\Diamond$. *For all $k_1 \geq 0$ there exist $k_2 \geq 0$ and $(v_1, v_2) \in W_1 \times W_2$ such that $S^{k_2}(w_2) \succcurlyeq v_2$, $v_1 \succcurlyeq S^{k_1}(w_1)$ and $v_1\,\mathcal{Z}_i\,v_2$.*

BACK $\Diamond$. *For all $k_2 \geq 0$ there exist $k_1 \geq 0$ and $(v_1, v_2) \in W_1 \times W_2$ such that $S^{k_1}(w_1) \succcurlyeq v_1$, $v_2 \succcurlyeq S^{k_2}(w_2)$ and $v_1\,\mathcal{Z}_i\,v_2$.*

The reader will notice that the clauses for $\Diamond$ involve the intuitionistic partial order, even though this is not involved in the semantics of $\Diamond$. However, this will give us more flexibility in designing bisimulations. The reason it works is that if k_1 is so that $S^{k_1}(w_1)$ witnesses that $\Diamond\varphi$ is true on w_1, then φ will also be true on any $v_1 \succcurlyeq S^{k_1}(w_1)$ by the monotonicity of intuitionistic truth. Similarly, if $S^{k_2}(w_2) \succcurlyeq v_2$ and φ holds on v_2, then it will also hold on $S^{k_2}(w_2)$. Thus we do not need $S^{k_1}(w_1)$ and $S^{k_2}(w_2)$ to be directly connected by the bisimulation; rather, it is sufficient for v_1, w_1 to act as 'proxies'. As was the case of Lemma 4.2, if two worlds are related by a bounded $\Diamond$-bisimulation, then they satisfy the same $\mathcal{L}_\Diamond$-formulas of small length.

Lemma 4.4. *Given two ITL^e models $\mathcal{M}_1$ and $\mathcal{M}_2$ and a bounded $\Diamond$-bisimulation $\mathcal{Z}_i \subseteq \cdots \subseteq \mathcal{Z}_0$ between them, for all $0 \leq i \leq n$ and $(w_1, w_2) \in W_1 \times W_2$, if $w_1\,\mathcal{Z}_i\,w_2$, then for all[3] $\varphi \in \mathcal{L}_\Diamond$ satisfying $|\varphi| \leq i$, $\mathcal{M}_1, w_1 \vDash \varphi$ iff $\mathcal{M}_2, w_2 \vDash \varphi$.*

Proof. We proceed by induction on n. Let $0 \leq i \leq n$ be such that for all $j < i$ the lemma holds. Let $w_1 \in W_1$ and $w_2 \in W_2$ be such that $w_1\,\mathcal{Z}_i\,w_2$ and let us consider $\varphi \in \mathcal{L}_\Diamond$ such that $|\varphi| \leq i$. We only consider the case where $\varphi = \Diamond\psi$, as other cases are covered by Lemma 4.2.

From left to right, if $\mathcal{M}_1, w_1 \vDash \Diamond\psi$ then there exists $k_1 \geq 0$ such that $\mathcal{M}_1, S^{k_1}(w_1) \vDash \psi$. By FORTH $\Diamond$, there exists $k_2 \geq 0$ and $(v_1, v_2) \in W_1 \times W_2$ such that $S^{k_2}(w_2) \succcurlyeq v_2$, $v_1 \succcurlyeq S^{k_1}(w_1)$ and $v_1\,\mathcal{Z}_{i-1}\,v_2$. By $\preccurlyeq$-monotonicity, $\mathcal{M}_1, v_1 \vDash \psi$. Then, by the induction hypothesis and the fact that $|\psi| \leq i - 1$, it follows that $\mathcal{M}_2, v_2 \vDash \psi$, thus by $\preccurlyeq$-monotonicity once again, $\mathcal{M}_2, S^{k_2}(w_2) \vDash \psi$, so that $\mathcal{M}_2, w_2 \vDash \Diamond\psi$. The converse direction is proved similarly by using BACK $\Diamond$. $\qquad\square$

We can define bounded $\Box$-bisimulations in a similar way.

---

[3]We remind the reader that, as per our convention, $\mathcal{L}_\Diamond$ is the $\Box, \mathcal{U}, \mathcal{R}$-free fragment. A similar comment applies to other sublanguages of $\mathcal{L}$ mentioned below.

Definition 4.5. *A bounded $\bigcirc$-bisimulation $\mathcal{Z}_n \subseteq \cdots \subseteq \mathcal{Z}_0 \subseteq W_1 \times W_2$ is said to be a bounded $\square$-bisimulation if for all $(w_1, w_2) \in W_1 \times W_2$ and for all $0 \le i < n$, if $w_1 \mathcal{Z}_{i+1} w_2$, then:*

FORTH $\square$. *For all $k_2 \ge 0$ there exist $k_1 \ge 0$ and $(v_1, v_2) \in W_1 \times W_2$ s.t. $S^{k_2}(w_2) \succcurlyeq v_2$, $v_1 \succcurlyeq S^{k_1}(w_1)$ and $v_1 \mathcal{Z}_i v_2$.*

BACK $\square$. *For all $k_1 \ge 0$ there exist $k_2 \ge 0$ and $(v_1, v_2) \in W_1 \times W_2$ s.t. $S^{k_1}(w_1) \succcurlyeq v_1$, $v_2 \succcurlyeq S^{k_2}(w_2)$ and $v_1 \mathcal{Z}_i v_2$.*

The intuition for the role of v_1, v_2 in the clauses for $\square$ is similar to that of $\diamond$, except that now we have to transfer *negative* information. If $\square\varphi$ *fails* at w_1, there will be $k_1 \ge 0$ so that φ fails on $S^{k_1}(w_1)$; but then, φ will forcibly fail on any $v_1 \preccurlyeq S^{k_1}(w_1)$. Similarly, if φ fails on $v_2 \succcurlyeq S^{k_2}(w_2)$, φ will fail on $S^{k_2}(w_2)$ as well, witnessing that $\square\varphi$ fails on w_2.

Lemma 4.6. *Given two ITL^e models $\mathcal{M}_1$ and $\mathcal{M}_2$ and a bounded $\square$-bisimulation $\mathcal{Z}_n \subseteq \cdots \subseteq \mathcal{Z}_0$ between them, for all $(w_1, w_2) \in W_1 \times W_2$ and $0 \le i \le n$, if $w_1 \mathcal{Z}_i w_2$ then for all $\varphi \in \mathcal{L}_\square$ such that $|\varphi| \le i$, then $\mathcal{M}_1, w_1 \vDash \varphi$ iff $\mathcal{M}_2, w_2 \vDash \varphi$.*

Proof. We proceed by induction on i. Let $i \ge 0$ be such that for all $j < i$ the lemma holds. Let $w_1 \in W_1$ and $w_2 \in W_2$ be such that $w_1 \mathcal{Z}_i w_2$ and let us consider $\varphi \in \mathcal{L}_\square$ such that $|\varphi| \le i$. Note that the cases for atoms as well as propositional and $\bigcirc$ connectives are proved as in Lemma 4.2, so we only consider $\varphi = \square\psi$.

For the left-to-right implication, we work by contrapositive, and assume that $\mathcal{M}_2, w_2 \nvDash \square\psi$. Then, there exists $k_2 \ge 0$ such that $\mathcal{M}_2, S^{k_2}(w_2) \nvDash \psi$. By FORTH $\square$, there exist $k_1 \ge 0$ and $(v_1, v_2) \in W_1 \times W_2$ s.t. $S^{k_2}(w_2) \succcurlyeq v_2$, $v_1 \succcurlyeq S^{i_1}(w_1)$ and $v_1 \mathcal{Z}_{i-1} v_2$. As in the proof of Lemma 4.4, by $\preccurlyeq$-monotonicity, the induction hypothesis and the fact that $|\psi| \le i-1$, it follows that $\mathcal{M}_1, v_1 \nvDash \psi$; thus $\mathcal{M}_1, S^{k_1}(w_1) \nvDash \psi$, and again by $\preccurlyeq$-monotonicity $\mathcal{M}_1, w_1 \nvDash \square\psi$. The converse direction follows a similar reasoning but using BACK $\square$. $\qquad\square$

5 Bounded bisimulations for $\mathcal{U}$ and $\mathcal{R}$

In this section we adapt the bisimulations defined for a language with *until* and *since* [18] presented by Kurtonina and de Rijke [20] to our case. As with bisimulations for $\diamond$ and $\square$, we modify the standard clauses so that witnesses for $\mathcal{U}$ or $\mathcal{R}$ do not have to be directly connected, and, instead, it suffices for suitable 'proxy' worlds to be connected by the bisimulation. Let us begin with bounded bisimulations for $\mathcal{U}$.

Definition 5.1. *Given* $n \in \mathbb{N}$ *and two* ITL^e *models* $\mathcal{M}_1$ *and* $\mathcal{M}_2$, *a bounded* $\circ$-*bisimulation* $\mathcal{Z}_n \subseteq \cdots \subseteq \mathcal{Z}_0 \subseteq W_1 \times W_2$ *is said to be a bounded* $\mathcal{U}$-*bisimulation iff for all* $(w_1, w_2) \in W_1 \times W_2$, *and for all* $0 \leq i < n$ *if* $w_1 \; \mathcal{Z}_{i+1} \; w_2$:

FORTH $\mathcal{U}$. *For all* $k_1 \geq 0$ *there exist* $k_2 \geq 0$ *and* $(v_1, v_2) \in W_1 \times W_2$ *such that*

1. $S^{k_2}(w_2) \succcurlyeq v_2$, $v_1 \succcurlyeq S^{k_1}(w_1)$ *and* $v_1 \; \mathcal{Z}_i \; v_2$, *and*

2. *for all* $j_2 \in [0, k_2)$ *there exist* $j_1 \in [0, k_1)$ *and* $(u_1, u_2) \in W_1 \times W_2$ *such that* $u_1 \succcurlyeq S^{j_1}(w_1)$, $S^{j_2}(w_2) \succcurlyeq u_2$ *and* $u_1 \; \mathcal{Z}_i \; u_2$.

BACK $\mathcal{U}$. *For all* $k_2 \geq 0$ *there exist* $k_1 \geq 0$ *and* $(v_1, v_2) \in W_1 \times W_2$ *such that*

1. $S^{k_1}(w_1) \succcurlyeq v_1$, $v_2 \succcurlyeq S^{k_2}(w_2)$ *and* $v_1 \; \mathcal{Z}_i \; v_2$, *and*

2. *for all* $j_1 \in [0, k_1)$ *there exist* $j_2 \in [0, k_2)$ *and* $(u_1, u_2) \in W_1 \times W_2$ *such that* $u_2 \succcurlyeq S^{j_2}(w_2)$, $S^{j_1}(w_1) \succcurlyeq u_1$ *and* $u_1 \; \mathcal{Z}_i \; u_2$.

As was the case before, the following lemma states that two bounded $\mathcal{U}$-bisimilar models agree on small $\mathcal{L}_{\mathcal{U}}$ formulas.

Lemma 5.2. *Given two* ITL^e *models* $\mathcal{M}_1$ *and* $\mathcal{M}_2$ *and a bounded* $\mathcal{U}$-*bisimulation* $\mathcal{Z}_n \subset \cdots \subset \mathcal{Z}_0$ *between them, for all* $0 \leq m \leq n$ *and* $(w_1, w_2) \in W_1 \times W_2$, *if* $w_1 \; \mathcal{Z}_m \; w_2$ *then for all* $\varphi \in \mathcal{L}_{\mathcal{U}}$ *such that* $|\varphi| \leq m$, $\mathcal{M}_1, w_1 \vDash \varphi$ *iff* $\mathcal{M}_2, w_2 \vDash \varphi$.

Proof. Once again, proceed by induction on n. Let $m \leq n$ be such that for all $k < m$ the lemma holds. Let $w_1 \in W_1$ and $w_2 \in W_2$ be such that $w_1 \; \mathcal{Z}_m \; w_2$ and let us consider $\varphi \in \mathcal{L}_{\mathcal{U}}$ such that $|\varphi| \leq m$. As before, we only consider the 'new' case, where $\varphi = \theta \, \mathcal{U} \, \psi$. From left to right, assume that $\mathcal{M}_1, w_1 \vDash \theta \, \mathcal{U} \, \psi$. Then, there exists $i_1 \geq 0$ such that $\mathcal{M}_1, S^{i_1}(w_1) \vDash \psi$ and for all j_1 satisfying $0 \leq j_1 < i_1$, $\mathcal{M}_1, S^{j_1}(w_1) \vDash \theta$. By FORTH $\mathcal{U}$, there exist $i_2 \geq 0$ and $(v_1, v_2) \in W_1 \times W_2$ such that 1. $S^{i_2}(w_2) \succcurlyeq v_2$, $v_1 \succcurlyeq S^{i_1}(w_1)$ and $v_1 \; \mathcal{Z}_{m-1} \; v_2$; 2. for all j_2 satisfying $0 \leq j_2 < i_2$ there exist $j_1 \in [0, i_1)$ and $(u_1, u_2) \in W_1 \times W_2$ s. t. $u_1 \succcurlyeq S^{j_1}(w_1)$, $S^{j_2}(w_2) \succcurlyeq u_2$ and $u_1 \; \mathcal{Z}_{m-1} \; u_2$.

From the first item, $\preccurlyeq$-monotonicity, the fact that $|\psi| \leq m - 1$, and the induction hypothesis, it follows that $\mathcal{M}_2, S^{i_2}(w_2) \vDash \psi$. Take any j_2 satisfying $0 \leq j_2 < i_2$. By the second item, the fact that $|\theta| \leq m - 1$, and the induction hypothesis, we conclude that $\mathcal{M}_2, S^{j_2}(w_2) \vDash \theta$ so $\mathcal{M}_2, w_2 \vDash \theta \, \mathcal{U} \, \psi$. The right-to-left direction is symmetric (but using BACK $\mathcal{U}$). $\square$

Finally, we define bounded bisimulations for $\mathcal{R}$.

Definition 5.3. *A bounded* $\circ$-*bisimulation* $\mathcal{Z}_n \subseteq \cdots \subseteq \mathcal{Z}_0 \subseteq W_1 \times W_2$ *is said to be a bounded* $\mathcal{R}$-*bisimulation if for all* $(w_1, w_2) \in W_1 \times W_2$ *and for all* $0 \leq i < n$, *if* $w_1 \; \mathcal{Z}_{i+1} \; w_2$ *then* :

FORTH $\mathcal{R}$. *For all* $k_2 \geq 0$ *there exist* $k_1 \geq 0$ *and* $(v_1, v_2) \in W_1 \times W_2$ *such that*

1. $S^{k_2}(w_2) \succcurlyeq v_2$, $v_1 \succcurlyeq S^{k_1}(w_1)$ and $v_1 \mathcal{Z}_i v_2$, and

2. for all j_1 satisfying $0 \le j_1 < k_1$ there exist j_2 such that $0 \le j_2 < k_2$ and $(u_1, u_2) \in W_1 \times W_2$ s. t. $u_1 \succcurlyeq S^{j_1}(w_1)$, $S^{j_2}(w_2) \succcurlyeq u_2$ and $u_1 \mathcal{Z}_i u_2$.

BACK $\mathcal{R}$. For all $k_1 \ge 0$ there exist $k_2 \ge 0$ and $(v_1, v_2) \in W_1 \times W_2$ such that

1. $S^{k_1}(w_1) \succcurlyeq v_1$, $v_2 \succcurlyeq S^{k_2}(w_2)$ and $v_1 \mathcal{Z}_i v_2$, and

2. for all j_2 satisfying $0 \le j_2 < k_2$ there exist j_1 such that $0 \le j_1 < k_1$ and $(u_1, u_2) \in W_1 \times W_2$ s. t. $u_2 \succcurlyeq S^{j_2}(w_2)$, $S^{j_1}(w_1) \succcurlyeq u_1$ and $u_1 \mathcal{Z}_i u_2$.

Once again, we obtain a corresponding bisimulation lemma for $\mathcal{L}_\mathcal{R}$.

Lemma 5.4. *Given two* ITL^e *models* $\mathcal{M}_1$ *and* $\mathcal{M}_2$ *and a bounded* $\mathcal{R}$-*bisimulation* $\mathcal{Z}_n \subseteq \cdots \subseteq \mathcal{Z}_0$ *between them, for all* $0 \le m \le n$ *and* $(w_1, w_2) \in W_1 \times W_2$, *if* $w_1 \mathcal{Z}_m w_2$ *then for all* $\varphi \in \mathcal{L}_\mathcal{U}$ *such that* $|\varphi| \le m$, $\mathcal{M}_1, w_1 \vDash \varphi$ *iff* $\mathcal{M}_2, w_2 \vDash \varphi$.

Proof. As before, we proceed by induction on n; the critical case where $\varphi = \theta \mathcal{R} \psi$ follows by a combination of the reasoning for Lemmas 4.6 and Lemma 4.6. Details are left to the reader. $\square$

6 Definability and undefinability of modal operators

In this section, we explore the question of when it is that the basic connectives can or cannot be defined in terms of each other. It is known that, classically, $\Diamond$ and $\square$ are interdefinable, as are $\mathcal{U}$ and $\mathcal{R}$; we will see that this is not the case intuitionistically. On the other hand, $\mathcal{U}$ (and hence $\mathcal{R}$) is not definable in terms of $\Diamond, \square$ in the classical setting [18], and this result immediately carries over to the intuitionistic setting, as the class of classical LTL models can be seen as the subclass of that of dynamic posets where the partial order is the identity.

Interdefinability of modal operators can vary within intermediate logics. For example, $\wedge$, $\vee$ and $\rightarrow$ are basic connectives in propositional intuitionistic logic, but in the intermediate logic of here-and-there [15], $\wedge$ [1, 2] and $\rightarrow$ [1] are basic operators while $\vee$ is definable in terms of $\rightarrow$ and $\wedge$ [22]. In first-order here-and-there [21], the quantifier $\exists$ is definable in terms of $\forall$ and $\rightarrow$ [24] while $\forall$ is not definable in terms of the other operators. In the modal case, Simpson [30] shows that modal operators are not interdefinable in the logic IK and Balbiani and Diéguez [2] proved the same result for the linear time temporal extension of here-and-there. This last proof is adapted to show that modal operators are not definable in ITL^e. Note, however, that here we correct the claim of [2] stating that $\Diamond$ is not here-and-there definable in terms of $\square$.

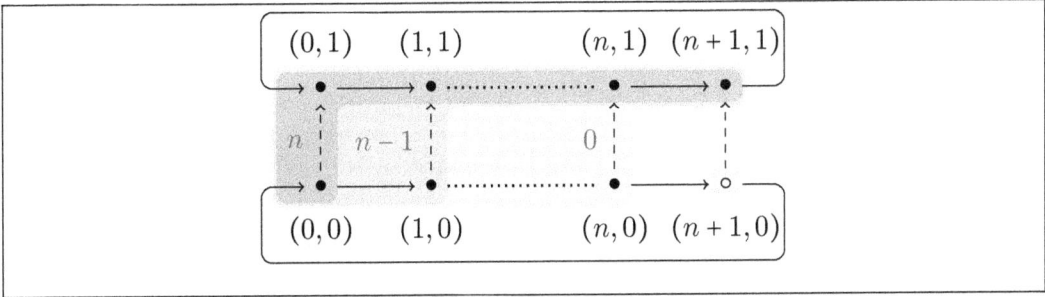

Figure 1: The here-and-there model $\mathcal{H}_n$. Black dots satisfy the atom p, white dots do not; all other atoms are false everywhere. Dashed lines indicate $\preccurlyeq$ and solid lines indicate S. The $\sim_i$-equivalence classes are shown as grey regions.

Let us begin by studying the definability of $\square$ in terms of $\circ$ and $\mathcal{U}$. Below, if $\mathcal{L}' \subseteq \mathcal{L}$, $\varphi \in \mathcal{L}$ and Ω is a class of models, we say that φ is $\mathcal{L}'$-definable over Ω if there is $\varphi' \in \mathcal{L}'$ such that $\Omega \vDash \varphi \leftrightarrow \varphi'$.

Theorem 6.1. *The connective $\square$ is not $\mathcal{L}_\mathcal{U}$-definable, even over the class of finite here-and-there models.*

Proof. Assume for the sake of contradiction that $\lozenge p$ can be expressed as a $\mathcal{U}$-free formula φ with $|\varphi| = n > 0$. Let $T = \{0, \cdots, n+1\}$ and $f : T \to T$ be given by $y = f(x)$ if and only if $y \equiv x + 1 \pmod{n+2}$. Then consider a here-and-there model $\mathcal{H}_n = (W, \preccurlyeq, S, V)$ based on $\mathrm{HT}(T, f)$ and with $V(p) = W \setminus \{(n+1, 0)\}$. For $k \leq n$, define $(i, j) \sim_k (i', j')$ if $(i, j) = (i', j')$ or

$$\max\{i(1-j), i'(1-j')\} \leq n - k$$

(see Figure 1). Clearly, $(\mathcal{H}_n, (0,0)) \not\vDash \lozenge p$, while $(\mathcal{H}_n, (0,1)) \vDash \lozenge p$. Let us check now that $(\sim_k)_{k \leq n}$ is a bounded $\mathcal{U}$-bisimulation. It is easy to check that the sequence is increasing under inclusion. Moreover, $\sim_k$ is symmetric (indeed, an equivalence relation) for eack k, so by symmetry, we only check the FORTH clauses.

ATOMS : Assume that $0 \leq k \leq n$ and $x \sim_k y$. Since $(n+1)(1-0) > n-k$, either $x = y$ (so the two satisfy the same atoms) or $x, y \neq (n+1, 0)$, so the two also satisfy the same atoms (namely, $\{p\}$).

FORTH $\to$: Let k satisfy $0 \leq k < n$ and let us assume $(i_1, j_1) \sim_{k+1} (i_2, j_2)$ and $(i_1, j_1) \preccurlyeq (i_1', j_1')$. If $(i_1, j_1) = (i_2, j_2)$, then $(i_2', j_2') \overset{def}{=} (i_1', j_1')$ witnesses that the clause holds, so we assume otherwise. Let us define $(i_2', j_2') \overset{def}{=} (i_2, 1)$. Then, $(i_2, j_2) \preccurlyeq (i_2', j_2')$ and $\max\{i_1'(1-j_1'), i_2'(1-j_2')\} = \max\{i_1'(1-j_1'), 0\} = i_1'(1-j_1') \leq n-k$, meaning that $(i_1', j_1') \sim_k (i_2', j_2')$, as required.

2279

FORTH $\bigcirc$: Let k satisfy $0 \le k < n$ and let us consider $(i_1, j_1) \sim_{k+1} (i_2, j_2)$. If $(i_1, j_1) = (i_2, j_2)$, then also $S(i_1, j_1) = S(i_2, j_2)$, so we assume otherwise. We claim that for $\ell \in \{1, 2\}$, $f(i_\ell)(1 - j_\ell) \le n - k$. If $j_\ell = 1$ this is obvious, otherwise from the definition of $\sim_{k+1}$ we obtain $i_\ell < n - k$ so that $f(i_\ell) = i_\ell + 1 \le n - k$. We conclude that $\max\{(f(i_1)(1 - j_1), f(i_2 + 1)(1 - j_2)\} \le n - k$, so that $S(i_1, j_1) \sim_k S(i_2, j_2)$, as required.

FORTH $\mathcal{U}$: Let k satisfy $0 \le k < n$, and let us suppose that $(i_1, j_1) \sim_{k+1} (i_2, j_2)$. Assume moreover that $(i_1, j_1) \ne (i_2, j_2)$, as the other case is easy to check. Fix $k_1 \ge 0$ and define $(i'_1, j'_1) = S^{k_1}(i_1, j_1)$. Let us define $k_2 = 0$, $v_1 = (i_1, 1)$, and $v_2 = (i_2, j_2)$, so that $S^{k_2}(i_2, j_2) = (i_2, j_2)$. Since $\max\{i_1(1 - 1), i_2(1 - j_2)\} = i_2(1 - j_2) < n - k$, we have that $v_1 \sim_k v_2$ and satisfy Condition 1. Note also that the Condition 2 holds vacuously because of $[0, k_2) = \varnothing$.

Consequently, $(\sim_m)_{m \le n}$ is a a bounded $\mathcal{U}$-bisimulation. By using Lemma 5.2 and the fact that $(0, 0) \sim_n (0, 1)$ we get that $(0, 0)$ and $(0, 1)$ satisfy the same $\mathcal{U}$-free formulas ψ with $|\psi| \le n$. However, $(\mathcal{H}_n, (0, 0)) \not\models \varphi$ and $(\mathcal{H}_n, (0, 1)) \models \varphi$: a contradiction. $\square$

As a consequence:

Corollary 6.2. *The connective* $\mathcal{R}$ *is not definable in terms of* $\bigcirc$ *and* $\mathcal{U}$, *even over the class of persistent models.*

Proof. If we could define $q \mathcal{R} p$, then we could also define $\square p \equiv \bot \mathcal{R} p$. $\square$

Proposition 6.3. *Over the class of here-and-there models,* $\Diamond$ *is* $\mathcal{L}_\square$*-definable. To be precise,* $\Diamond p$ *is equivalent to*

$$\varphi = (\square(p \to \square(p \vee \neg p)) \wedge \square(\bigcirc\square(p \vee \neg p) \to p \vee \neg p \vee \bigcirc\square\neg p)) \to (\square(p \vee \neg p) \wedge \neg\square\neg p).$$

Proof. Let $\mathcal{M} = (T \times \{0, 1\}, \preccurlyeq, S, V)$ be a here-and-there model with $S(t, i) = (f(t), i)$ (see Section 2.3). Before proving that φ is equivalent to $\Diamond p$, we give some intuition. Essentially, φ contemplates three different ways that $\Diamond p$ could hold in $(\mathcal{M}, x)$, where $x = (x_1, x_2)$. It may be that $\square(p \vee \neg p)$ holds, in which case $(\mathcal{M}, x)$ behaves essentially as a classical model, at least for formulas whose only variable is p. In this case, $\Diamond p$ holds iff $\neg\square\neg p$ holds, as in the standard classical semantics. If $\square(p \vee \neg p)$ fails, then $\mathcal{M}$ does not behave classically; for some n, $S^n(x)$ falsifies $p \vee \neg p$. For φ to be true, we then need for either $\square(p \to \square(p \vee \neg p))$ or $\square(\bigcirc\square(p \vee \neg p) \to p \vee \neg p \vee \bigcirc\square\neg p))$ to fail. The formula $\square(p \to \square(p \vee \neg p))$ will fail exactly when there is m such that $S^m(x)$ satisfies p (hence x satisfies $\Diamond p$), and $\mathcal{M}$ does *not* behave classically after m; that is, there is $n > m$ so that $S^n(x)$ falsifies $p \vee \neg p$. Meanwhile, $\square(\bigcirc\square(p \vee \neg p) \to$

$p \lor \neg p \lor \bigcirc \square \neg p))$ will fail exactly when there is m such that $S^m(x)$ satisfies p but $\mathcal{M}$ behaves classically after m; in other words, $S^n(x)$ falsifies $p \lor \neg p$ only for $n < m$. In this case, $\bigcirc \square (p \lor \neg p) \to p \lor \neg p \lor \bigcirc \square \neg p$ will be falsified exactly at the *greatest* such n.

Now for the proof. Assume that $x = (x_1, x_2)$ is such that $(\mathcal{M}, x) \vDash \Diamond p$. To check that $(\mathcal{M}, x) \vDash \varphi$, let $x' \succcurlyeq x$, so that $x' = (x_1, x_2')$ with $x_2' \geq x_2$, and consider the following cases.

CASE $(\mathcal{M}, x') \vDash \square(p \lor \neg p)$. In this case, it is easy to see that we also have $(\mathcal{M}, x') \vDash \neg \square \neg p$ given that $(\mathcal{M}, x) \vDash \Diamond p$.

CASE $(\mathcal{M}, x') \not\vDash \square(p \lor \neg p)$. Using the assumption that $(\mathcal{M}, x) \vDash \Diamond p$, choose k such that $(\mathcal{M}, (f^k(x_1), x_2)) \vDash p$ and consider two sub-cases.

1. Suppose there is $k' > k$ such that $(\mathcal{M}, (f^{k'}(x_1), x_2')) \not\vDash p \lor \neg p$. Then, it follows that $(\mathcal{M}, (f^k(x_1), x_2')) \not\vDash p \to \square p \lor \neg p$ and hence $(\mathcal{M}, x') \not\vDash \square(p \to \square(p \lor \neg p))$.

2. If there is not such k', then there must be a maximal $k' < k$ such that $(\mathcal{M}, (f^{k'}(x_1), x_2')) \not\vDash p \lor \neg p$ (otherwise, we would be in CASE $(\mathcal{M}, x') \vDash \square(p \lor \neg p)$). It is easily verified that

$$(\mathcal{M}, (f^{k'}(x_1), x_2')) \not\vDash \bigcirc \square(p \lor \neg p) \to p \lor \neg p \lor \bigcirc \square \neg p,$$

and hence

$$(\mathcal{M}, x') \not\vDash \square(\bigcirc \square(p \lor \neg p) \to p \lor \neg p \lor \bigcirc \square \neg p).$$

Note that the above direction does not use any properties of here-and-there models, and works over arbitrary expanding models. However, we need these properties for the other implication. Suppose that $(\mathcal{M}, x) \vDash \varphi$. If $(\mathcal{M}, x) \vDash \square(p \lor \neg p) \land \neg \square \neg p$, then it is readily verified that $(\mathcal{M}, x) \vDash \Diamond p$. Otherwise,

$$(\mathcal{M}, x) \not\vDash \square(p \to \square(p \lor \neg p)) \land \square(\bigcirc \square(p \lor \neg p) \to p \lor \neg p \lor \bigcirc \square \neg p).$$

If $(\mathcal{M}, x) \not\vDash \square(p \to \square(p \lor \neg p))$, then there is k such that

$$(\mathcal{M}, (f^k(x_1), x_2)) \not\vDash p \to \square(p \lor \neg p).$$

This is only possible if $x_2 = 0$ and $(\mathcal{M}, (f^k(x_1), x_2)) \vDash p$, so that $(\mathcal{M}, x) \vDash \Diamond p$. Similarly, if

$$(\mathcal{M}, x) \not\vDash \square(\bigcirc \square(p \lor \neg p) \to p \lor \neg p \lor \bigcirc \square \neg p),$$

then there is k such that $(\mathcal{M}, (f^k(x_1), x_2)) \not\vDash \bigcirc \square(p \lor \neg p) \to p \lor \neg p \lor \bigcirc \square \neg p$. This is only possible if $x_2 = 0$, $(\mathcal{M}, (f^k(x_1), x_2)) \vDash \bigcirc \square(p \lor \neg p)$ and $(\mathcal{M}, (f^k(x_1), x_2)) \not\vDash \bigcirc \square \neg p$. But from this it easily can be seen that there is $k' > k$ with $(\mathcal{M}, (f^{k'}(x_1), x_2)) \vDash p$, hence $(\mathcal{M}, x) \vDash \Diamond p$. $\qquad \square$

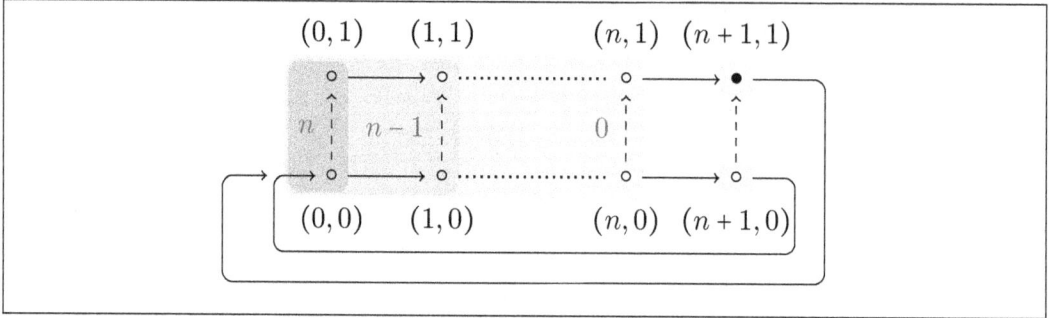

Figure 2: The expanding model $\mathcal{E}_n$. Notation is as in Figure 1.

Corollary 6.4. *Over the class of here-and-there models, $p\,\mathcal{U}\,q$ is $\mathcal{L}_\mathcal{R}$-definable using the equivalence $p\,\mathcal{U}\,q \equiv (q\,\mathcal{R}(p \vee q)) \wedge \Diamond q$.*

Hence, if we want to prove the undefinability of $\Diamond$ in terms of other operators, we must turn to a wider class of models, as we will do next.

Theorem 6.5. *The operator $\Diamond$ cannot be defined in terms of $\Box$ over the class of finite expanding models.*

Proof. Given $n > 0$, consider a model $\mathcal{E}_n = (W, \preccurlyeq, S, V)$ with $W = \{0, \cdots, n+1\} \times \{0, 1\}$, $(i, j) \preccurlyeq (i', j')$ if $i = i'$ and $j \leq j'$, $S(i, j) = (i+1, j)$ if $i \leq n$, $S(n+1, j) = (0, 0)$, and $V(p) = \{(n+1, 1)\}$. For $m \leq n$, define $(i, j) \sim_m (i', j')$ if either $(i, j) = (i', j')$, or $\max\{i, i'\} \leq n - m$. Then, it can easily be checked that $(\mathcal{M}, (0, 0)) \not\models \Diamond p$, $(\mathcal{M}, (0, 1)) \models \Diamond p$, and $(0, 0) \sim_m (0, 1)$.

It remains to check that $(\sim_m)_{m \leq n}$ is a bounded $\Box$-bisimulation. We focus on the $\Box$ clauses, and by symmetry, prove only BACK $\Box$. Suppose that $(i_1, j_1) \sim_m (i_2, j_2)$ and fix $k_1 \geq 0$. Let $(i'_1, j'_1) = S^{k_1}(i_1, j_1)$. Choose $k_2 > n + 1$ such that $i_2 + k_2 \equiv i'_1$ $(\mathrm{mod}\ n+1)$, and let $(i'_2, j'_2) = S^{k_2}(i_2, j_2)$. It is not hard to check that $i'_1 = i'_2$ and $j'_2 = 0$, from which we obtain $(i'_2, j'_2) \preccurlyeq (i'_1, j'_1)$. Hence, setting $v_1 = v_2 = (i'_2, j'_2)$ gives us the desired witnesses.

By letting n vary, we see that no $\mathcal{L}_\Box$-formula can be equivalent to $\Diamond p$. $\qquad\square$

7 Conclusions

In this paper we have investigated on ITL^e, an intuitionistic analogue of LTL based on expanding domain models from modal logic. We have shown that, as happens in other modal intuitionistic logics or modal intermediate logics, modal operators are not interdefinable.

Many open questions remain regarding intuitionistic temporal logics. We know that ITL^e is decidable, but the proposed decision procedure is non-elementary. However, there seems to be little reason to assume that this is optimal, raising the following question:

Question 7.1. *Are the satisfiability and validity problems for* ITL^e *elementary?*

Meanwhile, we saw in Theorems 2.2 and 2.4 that ITL^e has the strong finite model property, while ITL^p does not have the finite model property at all. However, it may yet be that ITL^p is decidable despite this.

Question 7.2. *Is* ITL^p *decidable?*

Regarding expressive completeness, it is known that LTL is expressively complete [18, 29, 12, 16]; there exists a one-to-one correspondence (over $\mathbb{N}$) between the temporal language and the monadic first-order logic equipped with a linear order and 'next' relation [12]. It is not known whether the same property holds between ITL^e and first-order intuitionistic logic.

Question 7.3. *Is $\mathcal{L}$ equally expressive to monadic first-order logic over the class of dynamic or persistent models?*

Finally, a sound and complete axiomatization for ITL^e remains to be found. The results we have presented here could be a first step in this direction, and we conclude with the following:

Question 7.4. *Are the* ITL^e*-valid formulas listed in this work, together with the intuitionistic tautologies and standard inference rules, complete for the class of dynamic posets? Is the logic augmented with $(\bigcirc p \to \bigcirc q) \to \bigcirc(p \to q)$ complete for the class of persistent models?*

References

[1] F. Aguado, P. Cabalar, D. Pearce, G. Pérez, and C. Vidal. A denotational semantics for equilibrium logic. *TPLP*, 15(4-5):620–634, 2015.

[2] P. Balbiani and M. Diéguez. Temporal here and there. In M. Loizos and A. Kakas, editors, *Logics in Artificial Intelligence*, pages 81–96. Springer, 2016.

[3] P. Blackburn, M. de Rijke, and Y. Venema. *Modal Logic*. Cambridge University Press, New York, NY, USA, 2001.

[4] P. Balbiani, J. Boudou, M. Diéguez, and D. Fernández-Duque. Intuitionistic Linear Temporal Logics. *ACM Trans. Comput. Log.* 21(2): 14:1–14:32 (2020)

[5] A.V. Chagrov and M. Zakharyaschev. *Modal Logic*, volume 35 of *Oxford logic guides*. Oxford University Press, 1997.

[6] D. Van Dalen. Intuitionistic logic. In *Handbook of Philosophical Logic*, volume 166, pages 225–339. Springer Netherlands, 1986.

[7] R. Davies. A temporal-logic approach to binding-time analysis. In *A Temporal-Logic Approach to Binding-Time Analysis*. LICS 1996: 184–195

[8] J.M. Davoren. On intuitionistic modal and tense logics and their classical companion logics: Topological semantics and bisimulations. *Annals of Pure and Applied Logic*, 161(3):349–367, 2009.

[9] W.B. Ewald. Intuitionistic tense and modal logic. *The Journal of Symbolic Logic*, 51(1):166–179, 1986.

[10] D. Fernández-Duque. The intuitionistic temporal logic of dynamical systems. *Log. Methods Comput. Sci.* 14(3) (2018)

[11] G Fischer Servi. Axiomatisations for some intuitionistic modal logics. In *Rend. Sem. Mat. Univers. Polit. Torino*, volume 42, pages 179–194, Torino, Italy, 1984.

[12] D. Gabbay, A. Pnueli, S. Shelah, and J. Stavi. On the Temporal Analysis of Fairness. In *POPL 1980*, 163–173,

[13] D. Gabelaia, A. Kurucz, F. Wolter, and M. Zakharyaschev. Non-primitive recursive decidability of products of modal logics with expanding domains. *Annals of Pure and Applied Logic*, 142(1-3):245–268, 2006.

[14] R. Goldblatt. *Logics of Time and Computation*. Number 7 in CSLI Lecture Notes. Center for the Study of Language and Information, Stanford, California, 1992. second edition.

[15] A. Heyting. *Die formalen Regeln der intuitionistischen Logik*. Sitzungsberichte der Preussischen Akademie der Wissenschaften. Physikalisch-mathematische Klasse. Deütsche Akademie der Wissenschaften zu Berlin, Mathematisch-Naturwissenschaftliche Klasse, 1930.

[16] I Hodkinson. Expressive completeness of Until and Since over Dedekind complete linear time In *Modal logic and process algebra*, ed. A. Ponse, M. de Rijke, Y. Venema, CSLI Lecture Notes 53, 1995, pp. 171–185.

[17] N. Kamide and H. Wansing. Combining linear-time temporal logic with constructiveness and paraconsistency. *J. Applied Logic*, 8(1):33–61, 2010.

[18] H. Kamp. *Tense Logic and the Theory of Linear Order*. PhD thesis, University of California, Los Angeles, California, USA, 1968.

[19] K. Kojima and A. Igarashi. Constructive linear-time temporal logic: Proof systems and Kripke semantics. *Information and Computation*, 209(12):1491 – 1503, 2011.

[20] N. Kurtonina and M. de Rijke. Bisimulations for temporal logic. *Journal of Logic, Language and Information*, 6(4):403–425, 1997.

[21] V. Lifschitz, D. Pearce, and A. Valverde. *A Characterization of Strong Equivalence for Logic Programs with Variables*, page 188–200. Springer Berlin Heidelberg, Berlin, Heidelberg, 2007.

[22] J. Łukasiewicz. Die Logik und das Grundlagenproblem. *Les Entretiens de Zürich sur les Fondements et la Méthode des Sciences Mathématiques*, 12(6-9):82–100, 1938.

[23] G. Mints. *A Short Introduction to Intuitionistic Logic.* Springer, 2000.

[24] Grigori Mints. Cut-free formulations for a quantified logic of here and there. *Annals of Pure and Applied Logic,* 162(3):237–242, 2010.

[25] H. Nishimura. Semantical analysis of constructive PDL. *Publications of the Research Institute for Mathematical Sciences, Kyoto University,* 18:427–438, 1982.

[26] A. Patterson. *Bisimulation and propositional intuitionistic logic,* page 347–360. Springer Berlin Heidelberg, Berlin, Heidelberg, 1997.

[27] G. Plotkin and C. Stirling. A Framework for Intuitionistic Modal Logics. *TARK 1986:* 399–406

[28] A. Pnueli. The Temporal Logic of Programs. *FOCS* 1977: 46-57

[29] A. Rabinovich. A Proof of Kamp's Theorem. *Logical Methods in Computer Science,* 10(1), 2014.

[30] A.K. Simpson. *The proof theory and semantics of intuitionistic modal logic.* PhD thesis, University of Edinburgh, UK, 1994.

 Received 15 October 2017

Intensionality, Intensional Recursion, and the Gödel-Löb Axiom

G. A. Kavvos

Department of Computer Science, University of Oxford
Wolfson Building, Parks Road, Oxford OX1 3QD, United Kingdom
`alex.kavvos@bristol.ac.uk`

Abstract

The use of a necessity modality in a typed λ-calculus can be used to separate it into two regions. These can be thought of as intensional vs. extensional data: data in the first region, the modal one, are available as code, and their description can be examined. In contrast, data in the second region are only available as values up to ordinary equality. This allows us to add non-functional operations at modal types whilst maintaining consistency. In this setting, the Gödel-Löb axiom acquires a novel constructive reading: it affords the programmer the possibility of a very strong kind of recursion which enables them to write programs that have access to their own code. This is a type of computational reflection that is strongly reminiscent of Kleene's Second Recursion Theorem.

1 Introduction

This paper is about putting a logical twist on two old pieces of programming lore:

- First, it is about using *modal types* to treat *programs-as-data* in a type-safe manner.

- Second, it is about noticing that—in the context of intensional programming— a constructive reading of the Gödel-Löb axiom, i.e. $\Box(\Box A \to A) \to \Box A$, amounts to a strange kind of recursion, namely *intensional recursion*.

We will introduce a *typed λ-calculus with modal types* that supports both of these features. We will call it *Intensional PCF*, after the simply-typed λ-calculus with **Y** introduced by Scott [34] and Plotkin [32].

This is a revised version of the third chapter of [23], which is in turn based on a paper presented at the 7th Workshop on Intuitionistic Modal Logic and Applications (IMLA 2017).

1.1 Intensionality and Programs-as-data

To begin, we want to discuss our notion of *programs-as-data*. We mean it in a way that is considerably stronger than the higher-order functional programming with which we are already familiar, i.e. 'functions as first-class citizens.' In addition to that, our notion hints at a kind of *homoiconicity*, similar to the one present in the LISP family of languages. It refers to the ability given to a programmer to *quote* code, and carry it around as a datum; see [5] for an instance of that in LISP. This ability can be used for *metaprogramming*, which is the activity of writing programs that write other programs. Indeed, this is what LISP macros excel at [12], and what the *metaprogramming community* has been studying for a long time; see e.g. [37, 39]. Considering programs as data—but in an untyped manner—was also the central idea in the work of the *partial evaluation community*: see [19, 16, 17].

But we would like to go even further. In LISP, a program is able to process code by treating it as mere symbols, thereby disregarding its function and behaviour. This is what we call *intensionality*: an operation is *intensional* if it is *finer than equality*. This amounts to a kind of *non-functional computation*. That this may be done type-theoretically was suspected by Davies and Pfenning [31, 9], who introduced modal types to programming language theory. A system based on nominal techniques that fleshed out those ideas was presented by Nanevski [29]. The notions of intensional and extensional equality implicit in this system were studied using logical relations by Pfenning and Nanevski [30]. However, none of these papers studied whether the induced equational systems are consistent. We show that, no matter the intensional mechanism at use, modalities enable consistent intensional programming.

To our knowledge, this paper presents the first consistency proof for intensional programming.

1.2 Intensional Recursion

We also want to briefly explain what we mean by *intensional recursion*; a fuller discussion may be found in [1, 23]. Most modern programming languages support *extensional recursion*: in the body of a function definition, the programmer may make a finite number of calls to the definiendum itself. Operationally, this leads a function to examine its own values at a finite set of points at which it has hopefully already been defined. In the *untyped λ-calculus*, with $=_\beta$ standing for β-convertibility, this is modelled by the *First Recursion Theorem (FRT)* [4, §6.1]:

Theorem 1 (First Recursion Theorem). $\forall f \in \Lambda.\ \exists u \in \Lambda.\ u =_\beta fu.$

However, as Abramsky [1] notes, in the *intensional paradigm* we have described above a stronger kind of recursion is attainable. Instead of merely examining the result of a finite number of recursive calls, the definiendum can recursively have access to a *full copy of its own source code.* This is embodied in Kleene's *Second Recursion Theorem (SRT)* [24]. Here is a version of the SRT in the untyped λ-calculus, where $\ulcorner u \urcorner$ means 'the Gödel number of the term u' [4, §6.5, Thm. 6.5.9].

Theorem 2 (Second Recursion Theorem). $\forall f \in \Lambda. \ \exists u \in \Lambda. \ u =_\beta f \ulcorner u \urcorner.$

Kleene also proved the following, where Λ^0 is the set of closed λ-terms:

Theorem 3 (Existence of Interpreter). $\exists \mathbf{E} \in \Lambda^0. \ \forall M \in \Lambda^0. \ \mathbf{E} \ulcorner M \urcorner \rightarrow^* M$

It is not hard to see that, using Theorem 3, the SRT implies the FRT for closed terms: given $f \in \Lambda^0$ we let $F \stackrel{\text{def}}{=} \lambda y. \ f(\mathbf{E} \, y)$, so that the SRT applied to F yields a term u such that

$$u =_\beta F \ulcorner u \urcorner =_\beta f \left(\mathbf{E} \ulcorner u \urcorner \right) =_\beta f \, u$$

It is not at all evident whether the converse holds. This is because the SRT is a *first-order theorem* that is about diagonalisation, Gödel numbers and source code, whereas the FRT really is about *higher types:* see the discussion in [23, §2].

Hence, in the presence of intensional operations, the SRT affords us with a much stronger kind of recursion. In fact, it allows for a certain kind of *computational reflection,* or *reflective programming,* of the same kind envisaged by Brian Cantwell Smith [35]. But the programme of Smith's *reflective tower* involved a rather mysterious construction with unclear semantics [10, 42, 8], eventually leading to a theorem that—even in the presence of a mild reflective construct, the so-called **fexpr**—observational equivalence of programs collapses to α-conversion: see Wand [41]. Similar forays have also been attempted by the partial evaluation community: see [14, 15, 18].

We will use modalities to stop intension from flowing back into extension, so that the aforementioned theorem in [41]—which requires unrestricted quoting—will not apply. We will achieve reflection by internalising the SRT. Suppose that our terms are typed, and that $u : A$. Suppose as well that there is a type constructor $\square$, so that $\square A$ means 'code of type A.' Then certainly $\ulcorner u \urcorner : \square A$, and f is forced to have type $\square A \rightarrow A$. A logical reading of the SRT is then the following: for every $f : \square A \rightarrow A$, there exists a $u : A$ such that $u = f \ulcorner u \urcorner$. This corresponds to *Löb's rule* from *provability logic* [7], namely

$$\frac{\square A \rightarrow A}{A}$$

which is equivalent to adding the Gödel-Löb axiom to the logic. In fact, the punchline of this paper is that *the type of the Second Recursion Theorem is the Gödel-Löb axiom of provability logic.*

To our knowledge, this paper presents the first sound, type-safe attempt at reflective programming.

1.3 Prospectus

In §2 we will introduce the syntax of iPCF, and in §3 we will show that it satisfies basic metatheoretic properties. Following that, in section §4 we will add intensional operations to iPCF. By proving that the resulting notion of reduction is confluent, we will obtain consistency for the system. We then look at the computational behaviour of some important terms in §5, and conclude with two key examples of the new powerful features of our language in §6.

2 Introducing Intensional PCF

Intensional PCF (iPCF) is a typed λ-calculus with modal types. As discussed before, the modal types work in our favour by separating intension from extension, so that the latter does not leak into the former. Given the logical flavour of our previous work on categorical models of intensionality [22], we shall model the types of iPCF after the *constructive modal logic* S4, in the dual-context style pioneered by Pfenning and Davies [31, 9]. Let us seize this opportunity to remark that (a) there are also other ways to capture S4, for which see the survey [20], and that (b) dual-context formulations are not by any means limited to S4: they began in the context of *intuitionistic linear logic* [3], but have recently been shown to also encompass other modal logics: see [21].

iPCF is *not* related to the language Mini-ML that is introduced by [9]: that is a call-by-value, ML-like language, with ordinary call-by-value fixed points. In contrast, ours is a call-by-name language with a new kind of fixed point, namely intensional fixed points. These fixed points will afford the programmer the full power of *intensional recursion*. In logical terms they correspond to throwing the Gödel-Löb axiom $\Box(\Box A \to A) \to \Box A$ into S4. Modal logicians might object to this, as, in conjunction with the T axiom $\Box A \to A$, it will make every type inhabited. We remind them that a similar situation occurs in PCF, where the $\mathbf{Y}_A : (A \to A) \to A$ combinator allows one to write a term $\mathbf{Y}_A(\lambda x : A.\ x)$ at every type A. As in the study of PCF, we care less about the logic and more about the underlying computation: *it is the terms that matter, and the types are only there to stop basic type errors from happening.*

The syntax and the typing rules of iPCF may be found in Figure 1. These are largely the same as Pfenning and Davies' S4, save the addition of some constants (drawn from PCF), and a rule for intensional recursion. The introduction rule for the modality restricts terms under a box $(-)$ to those containing only modal variables, i.e. variables carrying only intensions or code, but never 'live values:'

$$\frac{\Delta\,;\cdot \vdash M : A}{\Delta\,;\Gamma \vdash \mathsf{box}\ M : \Box A}$$

There is also a rule for intensional recursion:

$$\frac{\Delta\,;z : \Box A \vdash M : A}{\Delta\,;\Gamma \vdash \mathsf{fix}\ z\ \mathsf{in}\ M : A}$$

This will be coupled with the reduction $\mathsf{fix}\ z\ \mathsf{in}\ M \longrightarrow M[\mathsf{box}\ (\mathsf{fix}\ z\ \mathsf{in}\ M)/z]$. This rule is actually just *Löb's rule* with a modal context, and including it in the Hilbert system of a (classical or intuitionistic) modal logic is equivalent to including the Gödel-Löb axiom: see [7] and [40]. Finally, let us record the fact that erasing the modality from the types appearing in either Löb's rule or the Gödel-Löb axiom yields the type of $\mathbf{Y}_A : (A \to A) \to A$, as a rule in the first case, or axiomatically internalised as a constant in the second (both variants exist in the literature: see [13] and [27]). A similar observation for a stronger form of the Löb axiom underlies the stream of work on *guarded recursion* [28, 6]; we recommend the survey [25] for a broad coverage of constructive modalities with a provability-like flavour.

3   Metatheory

iPCF satisfies the expected basic results: structural and cut rules are admissible. This is no surprise given its origin in the well-behaved Davies-Pfenning calculus. We assume the typical conventions for λ-calculi: terms are identified up to α-equivalence, for which we write $\equiv$, and substitution $[\cdot/\cdot]$ is defined in the ordinary, capture-avoiding manner. Bear in mind that we consider occurrences of u in N to be bound in $\mathsf{let}\ \mathsf{box}\ u \Leftarrow M\ \mathsf{in}\ N$. Contexts Γ, Δ are lists of type assignments $x : A$. Furthermore, we shall assume that whenever we write a judgement like $\Delta\,;\Gamma \vdash M : A$, then Δ and Γ are *disjoint*, in the sense that $\mathsf{Vars}\,(\Delta) \cap \mathsf{Vars}\,(\Gamma) = \emptyset$, where $\mathsf{Vars}\,(x_1 : A_1, \ldots, x_n : A_n) \stackrel{\text{def}}{=} \{x_1, \ldots, x_n\}$. We write Γ, Γ' for the concatenation of disjoint contexts. Finally, we sometimes write $\vdash M : A$ whenever $\cdot\,;\cdot \vdash M : A$.

Theorem 4 (Structural & Cut)**.** *The following rules are admissible in iPCF:*

1. *(Weakening)*

$$\frac{\Delta\,;\Gamma, \Gamma' \vdash M : A}{\Delta\,;\Gamma, x : A, \Gamma' \vdash M : A}$$

Ground Types G ::=   Nat | Bool

Types A, B ::= $G \mid A \to B \mid \Box A$

Terms M, N ::= $x \mid \lambda x{:}A.\ M \mid MN \mid$ box $M \mid$ let box $u \Leftarrow M$ in $N \mid$
                            $\widehat{n} \mid$ true $\mid$ false $\mid$ succ $\mid$ pred $\mid$ zero? $\mid \supset_G \mid$ fix z in M

Contexts Γ, Δ ::= $\cdot \mid \Gamma, x : A$

$$\frac{}{\Delta\,;\Gamma \vdash \widehat{n} : \mathsf{Nat}} \qquad\qquad \frac{}{\Delta\,;\Gamma \vdash b : \mathsf{Bool}}\ (b \in \{\mathsf{true, false}\})$$

$$\frac{}{\Delta\,;\Gamma \vdash \mathsf{zero?} : \mathsf{Nat} \to \mathsf{Bool}} \qquad \frac{}{\Delta\,;\Gamma \vdash f : \mathsf{Nat} \to \mathsf{Nat}}\ (f \in \{\mathsf{succ, pred}\})$$

$$\frac{}{\Delta\,;\Gamma \vdash \supset_G : \mathsf{Bool} \to G \to G \to G}$$

$$\frac{}{\Delta\,;\Gamma, x{:}A, \Gamma' \vdash x : A}\ (\mathsf{var}) \qquad\qquad \frac{}{\Delta, u{:}A, \Delta'\,;\Gamma \vdash u : A}\ (\Box\mathsf{var})$$

$$\frac{\Delta\,;\Gamma, x{:}A \vdash M : B}{\Delta\,;\Gamma \vdash \lambda x{:}A.\ M : A \to B}\ (\to \mathcal{I}) \qquad \frac{\Delta\,;\Gamma \vdash M : A \to B \quad \Delta\,;\Gamma \vdash N : A}{\Delta\,;\Gamma \vdash MN : B}\ (\to \mathcal{E})$$

$$\frac{\Delta\,;\cdot \vdash M : A}{\Delta\,;\Gamma \vdash \mathsf{box}\ M : \Box A}\ (\Box\mathcal{I}) \qquad \frac{\Delta\,;\Gamma \vdash M : \Box A \quad \Delta, u{:}A\,;\Gamma \vdash N : C}{\Delta\,;\Gamma \vdash \mathsf{let\ box}\ u \Leftarrow M\ \mathsf{in}\ N : C}\ (\Box\mathcal{E})$$

$$\frac{\Delta\,;z : \Box A \vdash M : A}{\Delta\,;\Gamma \vdash \mathsf{fix}\ z\ \mathsf{in}\ M : A}\ (\Box\mathsf{fix})$$

Figure 1: Syntax and Typing Rules for Intensional PCF

2. *(Exchange)*

$$\frac{\Delta \,;\Gamma, x : A, y : B, \Gamma' \vdash M : C}{\Delta \,;\Gamma, y : B, x : A, \Gamma' \vdash M : C}$$

3. *(Contraction)*

$$\frac{\Delta \,;\Gamma, x : A, y : A, \Gamma' \vdash M : A}{\Delta \,;\Gamma, w : A, \Gamma' \vdash M[w, w/x, y] : A}$$

4. *(Cut)*

$$\frac{\Delta \,;\Gamma \vdash N : A \qquad \Delta \,;\Gamma, x : A, \Gamma' \vdash M : A}{\Delta \,;\Gamma, \Gamma' \vdash M[N/x] : A}$$

Theorem 5 (Modal Structural & Cut). *The following rules are admissible:*

1. *(Modal Weakening)*

$$\frac{\Delta, \Delta' \,;\Gamma \vdash M : C}{\Delta, u : A, \Delta' \,;\Gamma \vdash M : C}$$

2. *(Modal Exchange)*

$$\frac{\Delta, x : A, y : B, \Delta' \,;\Gamma \vdash M : C}{\Delta, y : B, x : A, \Delta' \,;\Gamma \vdash M : C}$$

3. *(Modal Contraction)*

$$\frac{\Delta, x : A, y : A, \Delta' \,;\Gamma \vdash M : C}{\Delta, w : A, \Delta' \,;\Gamma \vdash M[w, w/x, y] : C}$$

4. *(Modal Cut)*

$$\frac{\Delta \,;\cdot \vdash N : A \quad \Delta, u : A, \Delta' \,;\Gamma \vdash M : C}{\Delta, \Delta' \,;\Gamma \vdash M[N/u] : C}$$

3.1 Free variables

In this section we prove a theorem regarding the occurrences of free variables in well-typed terms of iPCF. It turns out that, if a variable occurs free under a box $(-)$ construct, then it has to be in the modal context. This is the property that enforces that *intensions can only depend on intensions.*

Definition 1 (Free variables).

1. The *free variables* $\mathsf{fv}\,(M)$ of a term M are defined by induction on the structure of the term:

$$\mathsf{fv}\,(x) \stackrel{\text{def}}{=} \{x\} \qquad\qquad \mathsf{fv}\,(MN) \stackrel{\text{def}}{=} \mathsf{fv}\,(M) \cup \mathsf{fv}\,(N)$$

$$\mathsf{fv}\,(\lambda x : A.\ M) \stackrel{\text{def}}{=} \mathsf{fv}\,(M) - \{x\} \qquad \mathsf{fv}\,(\mathsf{box}\ M) \stackrel{\text{def}}{=} \mathsf{fv}\,(M)$$

$$\mathsf{fv}\,(\mathsf{fix}\ z\ \mathsf{in}\ M) \stackrel{\text{def}}{=} \mathsf{fv}\,(M) - \{z\}$$

as well as

$$\mathsf{fv}\,(\mathsf{let\ box}\ u \Leftarrow M\ \mathsf{in}\ N) \stackrel{\text{def}}{=} \mathsf{fv}\,(M) \cup (\mathsf{fv}\,(N) - \{u\})$$

and $\mathsf{fv}\,(c) \stackrel{\text{def}}{=} \emptyset$ for any constant c.

2. The *unboxed free variables* $\mathsf{fv}_0\,(M)$ of a term are those that do *not* occur under the scope of a $\mathsf{box}\ (-)$ or $\mathsf{fix}\ z\ \mathsf{in}\ (-)$ construct. They are formally defined by replacing the following clauses in the definition of $\mathsf{fv}\,(-)$:

$$\mathsf{fv}_0\,(\mathsf{box}\ M) \stackrel{\text{def}}{=} \emptyset \qquad\qquad \mathsf{fv}_0\,(\mathsf{fix}\ z\ \mathsf{in}\ M) \stackrel{\text{def}}{=} \emptyset$$

3. The *boxed free variables* $\mathsf{fv}_{\geq 1}\,(M)$ of a term M are those that *do* occur under the scope of a $\mathsf{box}\ (-)$ construct. They are formally defined by replacing the following clauses in the definition of $\mathsf{fv}\,(-)$:

$$\mathsf{fv}_{\geq 1}\,(x) \stackrel{\text{def}}{=} \emptyset \qquad\qquad \mathsf{fv}_{\geq 1}\,(\mathsf{box}\ M) \stackrel{\text{def}}{=} \mathsf{fv}\,(M)$$

$$\mathsf{fv}_{\geq 1}\,(\mathsf{fix}\ z\ \mathsf{in}\ M) \stackrel{\text{def}}{=} \mathsf{fv}\,(M) - \{z\}$$

Theorem 6 (Free variables).

1. *For every term M, $\mathsf{fv}\,(M) = \mathsf{fv}_0\,(M) \cup \mathsf{fv}_{\geq 1}\,(M)$.*

2. *If and $\Delta\,;\Gamma \vdash M : A$, then*

$$\mathsf{fv}_0\,(M) \subseteq \mathsf{Vars}\,(\Gamma) \cup \mathsf{Vars}\,(\Delta)$$
$$\mathsf{fv}_{\geq 1}\,(M) \subseteq \mathsf{Vars}\,(\Delta)$$

Proof.

1. Trivial induction on M.

2. By induction on the derivation of $\Delta\,;\Gamma \vdash M : A$.

$\square$

4 Consistency of Intensional Operations

In this section we shall prove that the modal types of iPCF enable us to consistently add intensional operations on the modal types. These are *non-functional operations on terms* which are not ordinarily definable because they violate equality. All we have to do is assume them as constants at modal types, define their behaviour by introducing a notion of reduction, and then prove that the compatible closure of this notion of reduction is confluent. A known corollary of confluence is that the equational theory induced by the reduction is *consistent*, i.e. does not equate all terms.

There is a caveat involving extension flowing into intension. That is: we need to exclude from consideration terms where a variable bound by a λ occurs under the scope of a box $(-)$ construct. These will never be well-typed, but—since we discuss types and reduction orthogonally—we also need to explicitly exclude them here too.

4.1 Adding intensionality

Davies and Pfenning [31] suggested that the $\Box$ modality can be used to signify intensionality. In fact, in [31, 9] they had prevented reductions from happening under box $(-)$ construct, " [...] since this would violate its intensional nature." But the truth is that neither of these presentations included any genuinely non-functional operations at modal types, and hence their only use was for homogeneous staged metaprogramming. Adding intensional, non-functional operations is a more difficult task. Intensional operations are dependent on *descriptions* and *intensions* rather than *values* and *extensions*. Hence, unlike reduction and evaluation, they cannot be blind to substitution. This is something that quickly came to light as soon as Nanevski [29] attempted to extend the system of Davies and Pfenning to allow 'intensional code analysis' using nominal techniques.

A similar task was also recently taken up by Gabbay and Nanevski [11], who attempted to add a construct is-app to the system of Davies and Pfenning, along with the reduction rules

$$\text{is-app (box } PQ) \longrightarrow \text{true}$$
$$\text{is-app (box } M) \longrightarrow \text{false} \qquad \text{if } M \text{ is not of the form } PQ$$

The function computed by is-app is truly intensional, as it depends solely on the syntactic structure of its argument: it merely checks if it syntactically is an application or not. As such, it can be considered a *criterion of intensionality*, albeit an extreme one: its definability conclusively confirms the presence of computation up to syntax.

Gabbay and Nanevski tried to justify the inclusion of is-app by producing denotational semantics for modal types in which the semantic domain $[\![\Box A]\!]$ directly involves the actual closed terms of type $\Box A$. However, something seems to have gone wrong with substitution. In fact, we believe that their proof of soundness is wrong: it is not hard to see that their semantics is not stable under the second of these two reductions: take M to be u, and let the semantic environment map u to an application PQ, and then notice that this leads to $[\![\text{true}]\!] = [\![\text{false}]\!]$. We can also see this in the fact that their notion of reduction is *not confluent*. Here is the relevant counterexample: we can reduce like this:

$$\text{let box } u \Leftarrow \text{box } (PQ) \text{ in is-app (box } u) \longrightarrow \text{is-app (box } PQ) \longrightarrow \text{true}$$

But we could have also reduced like that:

$$\text{let box } u \Leftarrow \text{box } (PQ) \text{ in is-app (box } u) \longrightarrow \text{let box } u \Leftarrow \text{box } (PQ) \text{ in false} \longrightarrow \text{false}$$

This example is easy to find if one tries to plough through a proof of confluence: it is very clearly *not* the case that $M \longrightarrow N$ implies $M[P/u] \longrightarrow N[P/u]$ if u is under a box $(-)$, exactly because of the presence of intensional operations such as is-app.

Perhaps the following idea is more workable: let us limit intensional operations to a chosen set of functions $f : \mathcal{T}(A) \to \mathcal{T}(B)$ from terms of type A to terms of type B, and then represent them in the language by a constant $\tilde{f}$, such that $\tilde{f}(\text{box } M) \longrightarrow \text{box } f(M)$. This set of functions would then be chosen so that they satisfy some sanity conditions. Since we want to have a let construct that allows us to substitute code for modal variables, the following general situation will occur: if $N \longrightarrow N'$, we have

$$\text{let box } u \Leftarrow \text{box } M \text{ in } N \longrightarrow N[M/u]$$

but also

$$\text{let box } u \Leftarrow \text{box } M \text{ in } N \longrightarrow \text{let box } u \Leftarrow \text{box } M \text{ in } N' \longrightarrow N'[M/u]$$

Thus, in order to have confluence, we need $N[M/u] \longrightarrow N'[M/u]$. This will only be the case for reductions of the form $\tilde{f}(\text{box } M) \to \text{box } f(M)$ if

$$f(N[M/u]) \equiv f(N)[M/u]$$

i.e. if f is *substitutive*. But then a simple naturality argument gives that $f(N) \equiv f(u[N/u]) \equiv f(u)[N/u]$, and hence $\tilde{f}$ is already definable by

$$\lambda x : \Box A. \text{ let box } u \Leftarrow x \text{ in box } f(u)$$

so such a 'substitutive' function is not intensional after all.

In fact, the only truly intensional operations we can add to our calculus will be those acting on *closed* terms. We will see that this circumvents the problems that arise when intensionality interacts with substitution. Hence, we will limit intensional operations to the following set:

Definition 2 (Intensional operations). Let $\mathcal{T}_0(A)$ be the set of (α-equivalence classes of) closed terms M such that $\cdot\,;\cdot \vdash M : A$. Then, the set of *intensional operations*, $\mathcal{F}(A, B)$, is defined to be the set of all functions $f : \mathcal{T}_0(A) \to \mathcal{T}_0(B)$.

We will include all of these intensional operations $f : \mathcal{T}_0(A) \to \mathcal{T}_0(B)$ in our calculus as constants:

$$\overline{\Delta\,;\Gamma \vdash \tilde{f} : \Box A \to \Box B}$$

with reduction rule $\tilde{f}(\mathsf{box}\ M) \to \mathsf{box}\ f(M)$, under the proviso that M is closed. Of course, these also includes operations on terms that *might not be computable*. However, we are interested in proving consistency of intensional operations in the most general setting. The questions of which intensional operations are computable, and which primitives or mechanisms can and should be used to express them, are beyond the scope of this paper, and largely still open.

4.2   Reduction and Confluence

We introduce a notion of reduction for iPCF, which we present in Figure 2. Unlike many studies of PCF-inspired languages, we do not consider a reduction strategy but ordinary 'non-deterministic' β-reduction. We do so because are trying to show consistency of the induced equational theory.

The equational theory induced by this notion of reduction is a symmetric version of it, annotated with types. It is easy to write down, so we omit it. Note the fact that, like the calculus of Davies and Pfenning, we do *not* include the following congruence rule for the modality:

$$\frac{\Delta\,;\cdot \vdash M = N : A}{\Delta\,;\Gamma \vdash \mathsf{box}\ M = \mathsf{box}\ N : \Box A}\ (\Box\mathsf{cong})$$

In fact, the very absence of this rule is what will allow modal types to become intensional. Otherwise, the only new rules are intensional recursion, embodied by the rule ($\Box\mathsf{fix}$), and intensional operations, exemplified by the rule ($\Box\mathsf{int}$).

We note that it seems perfectly reasonable to think that we should allow reductions under fix, i.e. admit the rule

$$\frac{M \longrightarrow N}{\mathsf{fix}\ z\ \mathsf{in}\ M \longrightarrow \mathsf{fix}\ z\ \mathsf{in}\ N}$$

$$\frac{}{(\lambda x : A.\ M)N \longrightarrow M[N/x]}\ (\longrightarrow \beta) \qquad \frac{M \longrightarrow N}{\lambda x : A.\ M \longrightarrow \lambda x : A.\ N}\ (\text{cong}_\lambda)$$

$$\frac{M \longrightarrow N}{MP \longrightarrow NP}\ (\text{app}_1) \qquad\qquad \frac{P \longrightarrow Q}{MP \longrightarrow MQ}\ (\text{app}_2)$$

$$\frac{}{\text{let box } u \Leftarrow \text{box } M \text{ in } N \longrightarrow N[M/u]}\ (\Box\beta)$$

$$\frac{}{\text{fix } z \text{ in } M \longrightarrow M[\text{box } (\text{fix } z \text{ in } M)/z]}\ (\Box\text{fix})$$

$$\frac{M \text{ closed, } M \in \text{dom}(f)}{\tilde{f}(\text{box } M) \longrightarrow \text{box } f(M)}\ (\Box\text{int})$$

$$\frac{M \longrightarrow N}{\text{let box } u \Leftarrow M \text{ in } P \longrightarrow \text{let box } u \Leftarrow N \text{ in } P}\ (\text{let-cong}_1)$$

$$\frac{P \longrightarrow Q}{\text{let box } u \Leftarrow M \text{ in } P \longrightarrow \text{let box } u \Leftarrow M \text{ in } Q}\ (\text{let-cong}_2)$$

$$\frac{}{\text{zero? } \widehat{0} \longrightarrow \text{true}}\ (\text{zero?}_1) \qquad \frac{}{\text{zero? } \widehat{n+1} \longrightarrow \text{false}}\ (\text{zero?}_2)$$

$$\frac{}{\text{succ } \widehat{n} \longrightarrow \widehat{n+1}}\ (\text{succ}) \qquad \frac{}{\text{pred } \widehat{n} \longrightarrow \widehat{n \div 1}}\ (\text{pred})$$

$$\frac{}{\supset_G \text{ true } M\ N \longrightarrow M}\ (\supset_1) \qquad \frac{}{\supset_G \text{ false } M\ N \longrightarrow N}\ (\supset_2)$$

Figure 2: Reduction for Intensional PCF

as M and N are expected to be of type A, which need not be modal. However, the reduction fix z in $M \longrightarrow M[\text{box}\,(\text{fix } z \text{ in } M)/z]$ 'freezes' M under an occurrence of box $(-)$, so that no further reductions can take place within it. Thus, the above rule would violate the intensional nature of boxes. We were likewise compelled to define $\text{fv}_0\,(\text{fix } z \text{ in } M) \overset{\text{def}}{=} \emptyset$ in the previous section: we should already consider M to be intensional, or under a box.

We can now show that

Theorem 7. *The reduction relation* $\longrightarrow$ *is confluent.*

The easiest route to that theorem is to use a proof like that in [21], i.e. the method of *parallel reduction*. This kind of proof was originally discovered by Tait and Martin-Löf, and is nicely documented in [38]. Because of the intensional nature of our box $(-)$ constructs, ours will be more nuanced and fiddly. The proof can of course be skipped on a first reading.

Proof of confluence   We will use a variant of the proof in [21], i.e. the method of *parallel reduction*. This kind of proof was originally discovered by Tait and Martin-Löf, and is nicely documented in [38]. Because of the intensional nature of our box $(-)$ constructs, ours will be more nuanced and fiddly than any in *op. cit.* The method is this: we will introduce a second notion of reduction,

$$\Longrightarrow \,\subseteq \Lambda \times \Lambda$$

which we will 'sandwich' between reduction proper and its transitive closure:

$$\longrightarrow \,\subseteq\, \Longrightarrow \,\subseteq\, \longrightarrow^*$$

We will then show that $\Longrightarrow$ has the diamond property. By the above inclusions, the transitive closure $\Longrightarrow^*$ of $\Longrightarrow$ is then equal to $\longrightarrow^*$, and hence $\longrightarrow$ is Church-Rosser.

In fact, we will follow [38] in doing something better: we will define for each term M its *complete development*, $M^\star$. The complete development is intuitively defined by 'unrolling' all the redexes of M at once. We will then show that if $M \Longrightarrow N$, then $N \Longrightarrow M^\star$. $M^\star$ will then suffice to close the diamond:

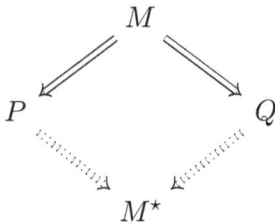

$$
\begin{array}{ccc}
 & M & \\
\swarrow & & \searrow \\
P & & Q \\
\searrow & & \swarrow \\
 & M^\star &
\end{array}
$$

$$\frac{}{M \Longrightarrow M} \text{ (refl)} \qquad \frac{M \Longrightarrow N \quad P \Longrightarrow Q}{(\lambda x : A.\ M)P \Longrightarrow N[Q/x]} (\to \beta)$$

$$\frac{M \Longrightarrow N}{\lambda x : A.\ M \Longrightarrow \lambda x : A.\ N} \text{ (cong}_\lambda) \qquad \frac{M \Longrightarrow N \quad P \Longrightarrow Q}{MP \Longrightarrow NQ} \text{ (app)}$$

$$\frac{P \Longrightarrow P'}{\supset_G \text{ true } P\ Q \Longrightarrow P'} (\supset_1) \qquad \frac{Q \Longrightarrow Q'}{\supset_G \text{ false } P\ Q \Longrightarrow Q'} (\supset_2)$$

$$\frac{M \Longrightarrow N}{\text{let box } u \Leftarrow \text{box } P \text{ in } M \Longrightarrow N[P/u]} (\Box\beta)$$

$$\frac{M \Longrightarrow N}{\text{fix } z \text{ in } M \Longrightarrow N[\text{box (fix } z \text{ in } M)/z]} (\Box\text{fix})$$

$$\frac{M \text{ closed}, M \in \text{dom}(f)}{\tilde{f}(\text{box } M) \Longrightarrow \text{box } f(M)} (\Box\text{int})$$

$$\frac{M \Longrightarrow N \quad P \Longrightarrow Q}{\text{let box } u \Leftarrow M \text{ in } P \Longrightarrow \text{let box } u \Leftarrow N \text{ in } Q} (\Box\text{let-cong})$$

Remark. In addition to the above, one should also include rules for the constants, but these are restatements of the rules in Figure 2.

Figure 3: Parallel Reduction

The parallel reduction $\Longrightarrow$ is defined in Figure 3. Instead of the axiom (refl) we would more commonly have an axiom for variables, $x \Longrightarrow x$, and $M \Longrightarrow M$ would be derivable. However, we do not have a congruence rule neither for box $(-)$ nor for Löb's rule, so that possibility would be precluded. We are thus forced to include $M \Longrightarrow M$, which slightly complicates the lemmas that follow.

The main lemma that usually underpins the confluence proof is this: if $M \Longrightarrow N$ and $P \Longrightarrow Q$, $M[P/x] \Longrightarrow N[Q/x]$. However, this is intuitively wrong: no reductions should happen under boxes, so this should only hold if we are substituting for a variable *not* occurring under boxes. Hence, this lemma splits into three different ones:

- $P \Longrightarrow Q$ implies $M[P/x] \Longrightarrow M[Q/x]$, if x does not occur under boxes: this is the price to pay for replacing the variable axiom with (refl).

- $M \implies N$ implies $M[P/u] \implies N[P/u]$, even if u is under a box.

- If x does not occur under boxes, $M \implies N$ and $P \implies Q$ indeed imply $M[P/x] \implies N[Q/x]$

Lemma 1. *If $M \implies N$ then $M[P/u] \implies N[P/u]$.*

Proof. By induction on the generation of $M \implies N$. Most cases trivially follow, or consist of simple invocations of the IH. In the case of $(\to \beta)$, the known substitution lemma suffices. Let us look at the cases involving boxes.

CASE($\Box\beta$). Then $M \implies N$ is let box $v \Leftarrow$ box R in $S \implies S'[R/v]$ with $S \implies S'$. By the IH, we have that $S[P/u] \implies S'[P/u]$, so

$$\text{let box } v \Leftarrow \text{box } R[P/u] \text{ in } S[P/u] \implies S'[P/u][R[P/u]/v]$$

and this last is α-equivalent to $S'[R/v][P/u]$ by the substitution lemma.

CASE($\Box$fix). A similar application of the substitution lemma.

CASE($\Box$int). Then $M \implies N$ is $\tilde{f}(\text{box } Q) \implies$ box $f(Q)$. Hence

$$\left(\tilde{f}(\text{box } Q) \right) [P/u] \equiv \tilde{f}(\text{box } Q) \implies \text{box } f(Q) \equiv (\text{box } f(Q)) [P/u]$$

simply because both Q and $f(Q)$ are closed.

$\square$

Lemma 2. *If $P \implies Q$ and $x \notin \mathsf{fv}_{\geq 1}(M)$, then $M[P/x] \implies M[Q/x]$.*

Proof. By induction on the term M. The only non-trivial cases are those for M a variable, box M' or fix z in M'. In the first case, depending on which variable M is, use either (refl), or the assumption $P \implies Q$. In the latter two, (box M')$[P/x] \equiv$ box $M' \equiv$ (box M')$[Q/x]$ as x does not occur under a box, so use (refl), and similarly for fix z in M'. $\square$

Lemma 3. *If $M \implies N$, $P \implies Q$, and $x \notin \mathsf{fv}_{\geq 1}(M)$, then*

$$M[P/x] \implies N[Q/x]$$

Proof. By induction on the generation of $M \implies N$. The cases for most congruence rules and constants follow trivially, or from the IH. We prove the rest.

CASE(refl). Then $M \Longrightarrow N$ is actually $M \Longrightarrow M$, so we use Lemma 2 to infer $M[P/x] \Longrightarrow M[Q/x]$.

CASE($\Box$int). Then $M \Longrightarrow N$ is actually $\tilde{f}(\text{box } M) \Longrightarrow \text{box } f(M)$. But M and $f(M)$ are closed, so $\left(\tilde{f}(\text{box } M) \right)[P/x] \equiv \tilde{f}(\text{box } M) \Longrightarrow \text{box } f(M) \equiv (\text{box } f(M))[Q/x]$.

CASE($\supset_i$). Then $M \Longrightarrow N$ is $\supset_G$ true $M\ N \Longrightarrow M'$ with $M \Longrightarrow M'$. By the IH, $M[P/x] \Longrightarrow M'[Q/x]$, so

$$\supset_G \text{ true } M[P/x]\ N[P/x] \Longrightarrow M'[Q/x]$$

by a single use of $(\supset_1)$. The case for false is similar.

CASE($\to \beta$). Then $(\lambda x'{:}A.\ M)N \Longrightarrow N'[M'/x']$, where $M \Longrightarrow M'$ and $N \Longrightarrow N'$. Then

$$((\lambda x'{:}A.\ M)N)\,[P/x] \equiv (\lambda x'{:}A.\ M[P/x])(N[P/x])$$

But, by the IH, $M[P/x] \Longrightarrow M'[Q/x]$ and $N[P/x] \Longrightarrow N'[Q/x]$. So by $(\to \beta)$ we have

$$(\lambda x'{:}A.\ M[P/x])(N[P/x]) \Longrightarrow M'[Q/x]\,[N'[Q/x]/x']$$

But this last is α-equivalent to $(M'[N'/x'])\,[Q/x]$ by the substitution lemma.

CASE($\Box\beta$). Then let box $u' \Leftarrow \text{box } M$ in $N \Longrightarrow N'[M/u']$ where $N \Longrightarrow N'$. By assumption, we have that $x \notin \text{fv}\,(M)$ and $x \notin \text{fv}_{\geq 1}\,(N)$. Hence, we have by the IH that $N[P/x] \Longrightarrow N'[Q/x]$, so by applying $(\Box\beta)$ we get

$$\begin{aligned}
(\text{let box } u' \Leftarrow \text{box } M \text{ in } N)[P/x] &\equiv \text{let box } u' \Leftarrow \text{box } M[P/x] \text{ in } N[P/x] \\
&\equiv \text{let box } u' \Leftarrow \text{box } M \text{ in } N[P/x] \\
&\Longrightarrow N'[Q/x][M/u']
\end{aligned}$$

But this last is α-equivalent to $N'[M/u'][Q/x]$, by the substitution lemma and the fact that x does not occur in M.

CASE($\Box$fix). Then fix z in $M \Longrightarrow M'[\text{box (fix } z \text{ in } M)/z]$, with $M \Longrightarrow M'$. As $x \notin \text{fv}_{\geq 1}\,(\text{fix } z \text{ in } M)$, we have that $x \notin \text{fv}\,(M)$, and by Lemma 5, $x \notin \text{fv}\,(M')$ either, so

$$(\text{fix } z \text{ in } M)[P/x] \equiv \text{fix } z \text{ in } M$$

and

$$M'[\text{fix } z \text{ in } M/z][Q/x] \equiv M'[Q/x][\text{fix } z \text{ in } M[Q/x]/z] \equiv M'[\text{fix } z \text{ in } M/z]$$

Thus, a single use of ($\Box$fix) suffices.

$\Box$

We now pull the following definition out of the hat:

Definition 3 (Complete development). The *complete development* $M^\star$ of a term M is defined by the following clauses:

$$x^\star \overset{\text{def}}{=} x$$
$$c^\star \overset{\text{def}}{=} c \qquad (c \in \{\tilde{f}, \widehat{n}, \text{zero?}, \dots\})$$
$$(\lambda x{:}A.\ M)^\star \overset{\text{def}}{=} \lambda x{:}A.\ M^\star$$
$$\left(\tilde{f}(\text{box } M)\right)^\star \overset{\text{def}}{=} \text{box } f(M) \qquad \text{if } M \text{ is closed}$$
$$((\lambda x{:}A.\ M)\ N)^\star \overset{\text{def}}{=} M^\star[N^\star/x]$$
$$(\supset_G\ \text{true } M\ N)^\star \overset{\text{def}}{=} M^\star$$
$$(\supset_G\ \text{false } M\ N)^\star \overset{\text{def}}{=} N^\star$$
$$(MN)^\star \overset{\text{def}}{=} M^\star N^\star$$
$$(\text{box } M)^\star \overset{\text{def}}{=} \text{box } M$$
$$(\text{let box } u \Leftarrow \text{box } M \text{ in } N)^\star \overset{\text{def}}{=} N^\star[M/u]$$
$$(\text{let box } u \Leftarrow M \text{ in } N)^\star \overset{\text{def}}{=} \text{let box } u \Leftarrow M^\star \text{ in } N^\star$$
$$(\text{fix } z \text{ in } M)^\star \overset{\text{def}}{=} M^\star[\text{box } (\text{fix } z \text{ in } M)/z]$$

We need the following two technical results as well.

Lemma 4. $M \Longrightarrow M^\star$

Proof. By induction on the term M. Most cases follow immediately by (refl), or by the IH and an application of the relevant rule. The case for box M follows by (refl), the case for fix z in M follows by ($\Box$fix), and the case for $\tilde{f}(\text{box } M)$ by ($\Box$int). $\Box$

Lemma 5 (BFV antimonotonicity). *If $M \Longrightarrow N$ then $\text{fv}_{\geq 1}(N) \subseteq \text{fv}_{\geq 1}(M)$.*

Proof. By induction on $M \Longrightarrow N$. $\Box$

And here is the main result:

Theorem 8. *If $M \Longrightarrow P$, then $P \Longrightarrow M^\star$.*

Proof. By induction on the generation of $M \Longrightarrow P$. The case of (refl) follows by Lemma 4, and the cases of congruence rules follow from the IH. We show the rest.

CASE($\to \beta$). Then we have $(\lambda x{:}A.\ M)N \Longrightarrow M'[N'/x]$, with $M \Longrightarrow M'$ and $N \Longrightarrow N'$. By the IH, $M' \Longrightarrow M^\star$ and $N' \Longrightarrow N^\star$. We have that $x \notin \mathsf{fv}_{\geq 1}(M)$, so by Lemma 5 we get that $x \notin \mathsf{fv}_{\geq 1}(M')$. Hence, by Lemma 3 we get $M'[N'/x] \Longrightarrow M^\star[N^\star/x] \equiv ((\lambda x{:}A.\ M)\,N)^\star$.

CASE($\square\beta$). Then we have

$$\mathsf{let\ box}\ u \Leftarrow \mathsf{box}\ M\ \mathsf{in}\ N \Longrightarrow N'[M/u]$$

where $N \Longrightarrow N'$. By the IH, $N' \Longrightarrow N^\star$, so it follows that

$$N'[M/u] \Longrightarrow N^\star[M/u] \equiv (\mathsf{let\ box}\ u \Leftarrow \mathsf{box}\ M\ \mathsf{in}\ N)^\star$$

by Lemma 1.

CASE($\square\mathsf{fix}$). Then we have

$$\mathsf{fix}\ z\ \mathsf{in}\ M \Longrightarrow M'[\mathsf{box}\ (\mathsf{fix}\ z\ \mathsf{in}\ M)/z]$$

where $M \Longrightarrow M'$. By the IH, $M' \Longrightarrow M^\star$. Hence

$$M'[\mathsf{box}\ (\mathsf{fix}\ z\ \mathsf{in}\ M)/z] \Longrightarrow M^\star[\mathsf{box}\ (\mathsf{fix}\ z\ \mathsf{in}\ M)/z] \equiv (\mathsf{fix}\ z\ \mathsf{in}\ M)^\star$$

by Lemma 1.

CASE($\square\mathsf{int}$). Similar.

$\square$

5   Some important terms

Let us look at the kinds of terms we can write in iPCF.

From the axioms of S4 First, we can write a term corresponding to axiom K, the *normality axiom* of modal logics:

$$\mathsf{ax}_\mathsf{K} \stackrel{\mathsf{def}}{=} \lambda f : \square(A \to B).\ \lambda x : \square A.\ \mathsf{let\ box}\ g \Leftarrow f\ \mathsf{in}\ \mathsf{let\ box}\ y \Leftarrow x\ \mathsf{in}\ \mathsf{box}\ (g\,y)$$

Then $\vdash \mathsf{ax_K} : \Box(A \to B) \to (\Box A \to \Box B)$. An intensional reading of this is the following: any function given as code can be transformed into an *effective operation* that maps code of type A to code of type B.

The rest of the axioms correspond to evaluating and quoting. Axiom T takes code to value, or intension to extension:

$$\vdash \mathsf{eval}_A \overset{\text{def}}{=} \lambda x : \Box A. \text{ let box } y \Leftarrow x \text{ in } y : \Box A \to A$$

and axiom $\mathsf{4}$ quotes code into code-for-code:

$$\vdash \mathsf{quote}_A \overset{\text{def}}{=} \lambda x : \Box A. \text{ let box } y \Leftarrow x \text{ in box } (\mathsf{box}\ y) : \Box A \to \Box\Box A$$

The Gödel-Löb axiom: intensional fixed points Since $(\Box\mathsf{fix})$ is Löb's rule, we expect to be able to write down a term corresponding to the Gödel-Löb axiom of provability logic. We can, and it is an *intensional fixed-point combinator*:

$$\mathbb{Y}_A \overset{\text{def}}{=} \lambda x : \Box(\Box A \to A). \text{ let box } f \Leftarrow x \text{ in box } (\mathsf{fix}\ z \text{ in } f\ z)$$

and $\vdash \mathbb{Y}_A : \Box(\Box A \to A) \to \Box A$. We observe that

$$\mathbb{Y}_A(\mathsf{box}\ M) \longrightarrow^* \mathsf{box}\ (\mathsf{fix}\ z \text{ in } (M\ z))$$

Undefined The combination of eval and intensional fixed points leads to non-termination, in a style reminiscent of the term $(\lambda x.\, xx)(\lambda x.\, xx)$ of the untyped λ-calculus. Let

$$\Omega_A \overset{\text{def}}{=} \mathsf{fix}\ z \text{ in } (\mathsf{eval}_A\ z)$$

Then $\vdash \Omega_A : A$, and

$$\Omega_A \longrightarrow \mathsf{eval}_A\ (\mathsf{box}\ \Omega_A) \longrightarrow^* \Omega_A$$

Extensional Fixed Points Perhaps surprisingly, the ordinary PCF $\mathbf{Y}$ combinator is also definable in the iPCF. Let

$$\mathbf{Y}_A \overset{\text{def}}{=} \mathsf{fix}\ z \text{ in } \lambda f : A \to A.\ f(\mathsf{eval}\ z\ f)$$

Then $\vdash \mathbf{Y}_A : (A \to A) \to A$, so that

$$\begin{aligned} \mathbf{Y}_A &\longrightarrow^* \lambda f : A \to A.\ f(\mathsf{eval}\ (\mathsf{box}\ \mathbf{Y}_A)\ f)) \\ &\longrightarrow^* \lambda f : A \to A.\ f(\mathbf{Y}_A\ f) \end{aligned}$$

6 Two intensional examples

No discussion of an intensional language with intensional recursion would be complete without examples that use these two novel features. Our first example uses intensionality, albeit in a 'extensional' way, and is drawn from the study of PCF and issues related to sequential vs. parallel (but not concurrent) computation. Our second example uses intensional recursion, so it is slightly more adventurous: it is a computer virus.

6.1 'Parallel or' by dovetailing

In [32] Gordon Plotkin proved the following theorem: there is no term $\mathsf{por} : \mathsf{Bool} \to \mathsf{Bool} \to \mathsf{Bool}$ of PCF such that $\mathsf{por}\ \mathsf{true}\ M \twoheadrightarrow_\beta \mathsf{true}$ and $\mathsf{por}\ M\ \mathsf{true} \twoheadrightarrow_\beta \mathsf{true}$ for any $\vdash M : \mathsf{Bool}$, whilst $\mathsf{por}\ \mathsf{false}\ \mathsf{false} \twoheadrightarrow_\beta \mathsf{false}$. Intuitively, the problem is that por has to first examine one of its two arguments, and this can be troublesome if that argument is non-terminating. It follows that the *parallel or* function is not definable in PCF. In order to regain the property of so-called *full abstraction* for the *Scott model* of PCF, a constant denoting this function has to be manually added to PCF, and endowed with the above rather clunky operational semantics. See [32, 13, 27, 36].

However, the parallel or function is a computable *partial recursive functional* [36, 26]. The way to prove that is intuitively the following: given two closed terms $M, N : \mathsf{Bool}$, take turns in β-reducing each one for a one step: this is called *dovetailing*. If at any point one of the two terms reduces to true, then output true. But if at any point both reduce to false, then output false.

This procedure is not definable in PCF because a candidate term por does not have access to a code for its argument, but can only inspect its value. However, in iPCF we can use the modality to obtain access to code, and intensional operations to implement reduction. Suppose we pick a reduction strategy $\longrightarrow_r$. Then, let us include a constant $\mathsf{tick} : \Box\mathsf{Bool} \to \Box\mathsf{Bool}$ that implements one step of this reduction strategy on closed terms:

$$\frac{M \longrightarrow_r N,\ M, N \text{ closed}}{\mathsf{tick}\ (\mathsf{box}\ M) \longrightarrow \mathsf{box}\ N}$$

Also, let us include a constant $\mathsf{done?} : \Box\mathsf{Bool} \to \mathsf{Bool}$, which tells us if a closed term under a box is a normal form:

$$\frac{M \text{ closed, normal}}{\mathsf{done?}\ (\mathsf{box}\ M) \longrightarrow \mathsf{true}} \qquad \frac{M \text{ closed, not normal}}{\mathsf{done?}\ (\mathsf{box}\ M) \longrightarrow \mathsf{false}}$$

These two can be subsumed under our previous scheme for introducing intensional operations. The above argument is now implemented by the following term:

$$\mathsf{por} :\equiv \mathbf{Y}(\lambda\,\mathsf{por}.\,\lambda x : \Box\mathsf{Bool}.\,\lambda y : \Box\mathsf{Bool}.$$
$$\supset_{\mathsf{Bool}} (\mathsf{done?}\ x)\ (\mathsf{lor}\ (\mathsf{eval}\ x)(\mathsf{eval}\ y))$$
$$(\supset_{\mathsf{Bool}} (\mathsf{done?}\ y) \quad (\mathsf{ror}\ (\mathsf{eval}\ x)(\mathsf{eval}\ y))$$
$$(\mathsf{por}\ (\mathsf{tick}\ x)(\mathsf{tick}\ y)))$$

where $\mathsf{lor}, \mathsf{ror} : \mathsf{Bool} \to \mathsf{Bool} \to \mathsf{Bool}$ are terms defining the left-strict and right-strict versions of the 'or' connective respectively. Notice that the type of this term is $\Box\mathsf{Bool} \to \Box\mathsf{Bool} \to \mathsf{Bool}$: we require *intensional access* to the terms of boolean type in order to define this function!

6.2   A computer virus

Abstract computer virology is the study of formalisms that model computer viruses. There are many ways to formalise viruses. We will use the model of Adleman [2], where files can be interpreted either as data, or as functions. We introduce a data type F of files, and two constants

$$\mathsf{in} : \Box(F \to F) \to F \quad \text{and} \quad \mathsf{out} : F \to \Box(F \to F)$$

If F is a file, then $\mathsf{out}\ F$ is that file interpreted as a program, and similarly for in. We ask that $\mathsf{out}\ (\mathsf{in}\ M) \longrightarrow M$, making $\Box(F \to F)$ a retract of F. This might seem the same as the situation where $F \to F$ is a retract of F, which yields models of the (untyped) λ-calculus, and is not trivial to construct [4, §5.4]. However, in our case it is not nearly as worrying: $\Box(F \to F)$ is populated by programs and codes, not by actual functions. Under this interpretation, the pair $(\mathsf{in}, \mathsf{out})$ corresponds to a kind of Gödel numbering—especially if F is $\mathbb{N}$.

In Adleman's model, a *virus* is given by its infected form, which either *injures*, *infects*, or *imitates* other programs. The details are unimportant in the present discussion, save from the fact that the virus needs to have access to code that it can use to infect other executables. One can hence construct such a virus from its *infection routine*, by using Kleene's SRT. Let us model it by a term

$$\vdash \mathsf{infect} : \Box(F \to F) \to F \to F$$

which accepts a piece of viral code and an executable file, and it returns either the file itself, or a version infected with the viral code. We can then define a term

$$\vdash \mathsf{virus} \stackrel{\mathrm{def}}{=} \mathsf{fix}\ z\ \mathsf{in}\ (\mathsf{infect}\ z) : F \to F$$

so that

$$virus \longrightarrow^* infect\ (box\ virus)$$

which is a program that is ready to infect its input with its own code.

7   Conclusion

We have achieved the desideratum of an intensional programming calculus with intensional recursion. There are two main questions that result from this development.

First, does there exist a good set of *intensional primitives* from which all others are definable? Is there perhaps *more than one such set*, hence providing us with a choice of programming primitives? Previous attempts aiming to answer this question include those of [33, 29].

Second, what is the exact kind of programming power that we have unleashed? Does it lead to interesting programs that we have not been able to write before? We have outlined some speculative applications for intensional recursion in [23, §§1–2]. Is iPCF a useful tool when it comes to attacking these?

Acknowledgements

The author would like to thank Mario Alvarez-Picallo for their endless conversations on types and metaprogramming, Neil Jones for his careful reading and helpful comments, and Samson Abramsky for suggesting the topic of intensionality. This work was supported by the EPSRC (award reference 1354534).

References

[1] Samson Abramsky. Intensionality, Definability and Computation. In Alexandru Baltag and Sonja Smets, editors, *Johan van Benthem on Logic and Information Dynamics*, pages 121–142. Springer-Verlag, 2014.

[2] Leonard M. Adleman. An Abstract Theory of Computer Viruses. In *Advances in Cryptology - CRYPTO' 88*, volume 403 of *Lecture Notes in Computer Science*, pages 354–374. Springer New York, New York, NY, 1990.

[3] Andrew Graham Barber. Dual Intuitionistic Linear Logic. Technical report, ECS-LFCS-96-347, Laboratory for Foundations of Computer Science, University of Edinburgh, 1996.

[4] Henk Barendregt. *Lambda Calculus: Its Syntax and Semantics*. North-Holland, Amsterdam, 1984.

[5] Alan Bawden. Quasiquotation in LISP. In *Proceedings of the 6th ACM SIGPLAN Workshop on Partial Evaluation and Semantics-Based Program Manipulation (PEPM '99)*, 1999.

[6] Lars Birkedal, Rasmus Møgelberg, Jan Schwinghammer, and Kristian Støvring. First steps in synthetic guarded domain theory: step-indexing in the topos of trees. *Logical Methods in Computer Science*, 8(4):1–45, oct 2012.

[7] George S. Boolos. *The Logic of Provability*. Cambridge University Press, Cambridge, 1994.

[8] Olivier Danvy and Karoline Malmkjaer. Intensions and extensions in a reflective tower. In *Proceedings of the 1988 ACM conference on LISP and functional programming (LFP '88)*, pages 327–341, New York, New York, USA, 1988. ACM Press.

[9] Rowan Davies and Frank Pfenning. A modal analysis of staged computation. *Journal of the ACM*, 48(3):555–604, 2001.

[10] Daniel P. Friedman and Mitchell Wand. Reification: Reflection without metaphysics. In *Proceedings of the 1984 ACM Symposium on LISP and functional programming (LFP '84)*, pages 348–355, New York, New York, USA, 1984. ACM Press.

[11] Murdoch J. Gabbay and Aleksandar Nanevski. Denotation of contextual modal type theory (CMTT): Syntax and meta-programming. *Journal of Applied Logic*, 11(1):1–29, mar 2013.

[12] Paul Graham. *On LISP: Advanced Techniques for Common LISP*. Prentice Hall, 1993.

[13] Carl A. Gunter. *Semantics of programming languages: structures and techniques*. Foundations of Computing. The MIT Press, 1992.

[14] Torben Amtoft Hansen, Thomas Nikolajsen, Jesper Larsson Träff, and Neil D. Jones. Experiments with Implementations of Two Theoretical Constructions. In *Proceedings of the Symposium on Logical Foundations of Computer Science: Logic at Botik '89*, pages 119–133, London, UK, 1989. Springer-Verlag.

[15] Neil D. Jones. Computer Implementation and Applications of Kleene's S-M-N and Recursion Theorems. In Yiannis N Moschovakis, editor, *Logic from Computer Science: Proceedings of a Workshop Held November 13-17, 1989 [at MSRI]*, volume 21 of *Mathematical Sciences Research Institute Publications*, pages 243–263. Springer New York, 1992.

[16] Neil D. Jones. An introduction to partial evaluation. *ACM Computing Surveys*, 28(3):480–503, 1996.

[17] Neil D. Jones. *Computability and Complexity: From a Programming Perspective*. Foundations of Computing. MIT Press, 1997.

[18] Neil D. Jones. A Swiss Pocket Knife for Computability. *Electronic Proceedings in Theoretical Computer Science*, 129:1–17, sep 2013.

[19] Neil D. Jones, Carsten K. Gomard, and Peter Sestoft. *Partial Evaluation and Automatic Program Generation*. Prentice Hall International, 1993.

[20] G. A. Kavvos. The Many Worlds of Modal λ-calculi: I. Curry-Howard for Necessity, Possibility and Time. *CoRR*, 2016.

[21] G. A. Kavvos. Dual-context calculi for modal logic. In *2017 32nd Annual ACM/IEEE Symposium on Logic in Computer Science (LICS)*. IEEE, 2017.

[22] G. A. Kavvos. On the Semantics of Intensionality. In Javier Esparza and Andrzej S. Murawski, editors, *Proceedings of the 20th International Conference on Foundations of Software Science and Computation Structures (FoSSaCS)*, volume 10203 of *Lecture Notes in Computer Science*, pages 550–566. Springer-Verlag Berlin Heidelberg, 2017.

[23] Georgios Alexandros Kavvos. *On the Semantics of Intensionality and Intensional Recursion*. DPhil thesis, University of Oxford, 2017.

[24] Stephen C. Kleene. On notation for ordinal numbers. *The Journal of Symbolic Logic*, 3(04):150–155, 1938.

[25] Tadeusz Litak. Constructive Modalities with Provability Smack. In Guram Bezhanishvili, editor, *Leo Esakia on duality in modal and intuitionistic logics*, pages 179–208. Springer, 2014.

[26] John R. Longley and Dag Normann. *Higher-Order Computability*. Theory and Applications of Computability. Springer Berlin Heidelberg, Berlin, Heidelberg, 2015.

[27] John C. Mitchell. *Foundations for programming languages*. Foundations of Computing. The MIT Press, 1996.

[28] Hiroshi Nakano. A modality for recursion. *Proceedings Fifteenth Annual IEEE Symposium on Logic in Computer Science (Cat. No.99CB36332)*, 2000.

[29] Aleksandar Nanevski. Meta-programming with names and necessity. *ACM SIGPLAN Notices*, 37:206–217, 2002.

[30] Aleksandar Nanevski and Frank Pfenning. Staged computation with names and necessity. *Journal of Functional Programming*, 15(06):893, 2005.

[31] Frank Pfenning and Rowan Davies. A judgmental reconstruction of modal logic. *Mathematical Structures in Computer Science*, 11(4):511–540, 2001.

[32] Gordon D. Plotkin. LCF considered as a programming language. *Theoretical Computer Science*, 5(3):223–255, 1977.

[33] Carsten Schürmann, Joëlle Despeyroux, and Frank Pfenning. Primitive recursion for higher-order abstract syntax. *Theoretical Computer Science*, 266(1-2):1–57, 2001.

[34] Dana S. Scott. A type-theoretical alternative to ISWIM, CUCH, OWHY. *Theoretical Computer Science*, 121(1-2):411–440, 1993.

[35] Brian Cantwell Smith. Reflection and Semantics in LISP. In *Proceedings of the 11th ACM SIGACT-SIGPLAN Symposium on Principles of Programming Languages (POPL '84)*, pages 23–35, New York, New York, USA, 1984. ACM Press.

[36] Thomas Streicher. *Domain-theoretic Foundations of Functional Programming*. World Scientific, 2006.

[37] Walid Taha and Tim Sheard. MetaML and multi-stage programming with explicit annotations. *Theoretical Computer Science*, 248(1-2):211–242, 2000.

[38] M. Takahashi. Parallel Reductions in λ-Calculus. *Information and Computation*, 118(1):120–127, apr 1995.

[39] Takeshi Tsukada and Atsushi Igarashi. A logical foundation for environment classifiers.

Logical Methods in Computer Science, 6(4):1–43, 2010.

[40] Aldo Ursini. A modal calculus analogous to K4W, based on intuitionistic propositional logic. *Studia Logica*, 38(3):297–311, 1979.

[41] Mitchell Wand. The Theory of Fexprs is Trivial. *LISP and Symbolic Computation*, 10(3):189–199, 1998.

[42] Mitchell Wand and Daniel P. Friedman. The mystery of the tower revealed: A nonreflective description of the reflective tower. *Lisp and Symbolic Computation*, 1(1):11–38, jun 1988.

    Received 14 October 2017

Justification Logic
for Constructive Modal Logic

Roman Kuznets*
TU Wien, Austria
roma@logic.at

Sonia Marin†
UCL, United Kingdom
s.marin@ucl.ac.uk

Lutz Straßburger
Inria, France
lutz@lix.polytechnique.fr

Abstract

We provide a treatment of the intuitionistic $\diamond$ modality in the style of justification logic. We introduce a new type of terms, called satisfiers, that justify consistency, obtain justification analogs for the constructive modal logics CK, CD, CT, and CS4, and prove the realization theorem for them.

1 Introduction

Justification logic is a family of modal logics generalizing the Logic of Proofs LP, introduced by Artemov in [6]. The original motivation, which was inspired by works of Kolmogorov and Gödel in the 1930's, was to give a classical provability semantics

We thank Björn Lellmann for helpful discussions and valuable input on the sequent calculi for constructive modal logics. We also thank anonymous reviewers for valuable comments. Travel for this collaboration was funded by the Austrian–French Scientific & Technological Cooperation Amadeus/Amadée grant "Analytic Calculi for Modal Logic" and by the ANR-FWF grant "FISP" (ANR-15-CE25-0014).

*Supported by the Austrian Science Fund (FWF) grants M 1770-N25, P 25417-G15, and ZK 35.
†Funded by the ERC Advanced Grant "ProofCert".

to intuitionistic propositional logic. Gödel [20] made the first steps by translating intuitionistic logic into the modal logic S4, which he rediscovered as a logic of abstract provability. He noted that S4-provability is incompatible with arithmetical reasoning due to the former's acceptance of the reflection principle and outlined, in an unpublished lecture [21], a potential way of overcoming this obstacle by descending to the level of proofs rather than provability. Artemov independently implemented essentially the same idea in the Logic of Proofs by showing that it provides an operational view of the same type of provability as S4 [6, 7].

The language of the Logic of Proofs can be seen as a modal language where occurrences of the $\Box$ modality are replaced with proof terms, also known as *proof polynomials*, *evidence terms*, or *justification terms*, depending on the setting. The intended meaning of the formula '$t : A$' is 't *is a proof of* A' or, more generally, the reason for the validity of A. Thus, the justification language is viewed as a refinement of the modal language, with one provability construct $\Box$ replaced with an infinite family of specific proofs.

It gradually became clear that the applicability of this result goes way beyond the provability interpretation of the modality, and can be equally well considered in other settings, including, notably, epistemic logic [9]. Indeed, the connection between the Logic of Proofs and the modal logic S4 has been extended to other modal logics (based on classical propositional reasoning), including normal modal sublogics of S4 [13], the modal logic S5 [11], all 15 logics of the so-called modal cube between the minimal normal modal logic K and S5 [22], the infinite family of Geach logics [18], to a certain extent to public announcement logic [14], etc. For more information on justification logic, the reader is referred to the entry [3] in the Stanford Encyclopedia of Philosophy, as well as to two recent books [4, 24] on the subject.

The correspondence between a justification logic and a modal logic means that erasing specific reasons in a valid statement about proofs leads to a valid statement about provability and, vice versa, any valid statement about provability can be viewed as a *forgetful projection* of a valid statement about proofs. Moreover, this existential view of $\Box$ as '*there exists a proof*' leads to a first-order provability reading of modal statements and suggests that they can be Skolemized. Such a Skolemization makes negative occurrences of $\Box$ into Skolem variables and positive occurrences into Skolem functions, suggesting a further restriction on the way the $\Box$ modalities are filled in with proof terms—the process called *realization*—negative occurrences should be filled in with distinct proof variables.

The Logic of Proofs was born out of an analysis of intuitionistic logic with the goal of explaining it using *classical reasoning* about proofs. However, other relationships with intuitionistic logic have also been explored. Artemov introduced the first intuitionistic version ILP of the Logic of Proofs in [8] to unify the semantics of modalities and lambda-calculus. Indeed, as simply typed lambda-calculus is in correspondence with intuitionistic proofs, he needed to define an intuitionistic axiomatization of the Logic of Proofs to relate modal logic S4 and λ-calculus. His axiomatization simply changes the propositional base to intuitionistic while keeping the other axioms of Logic of Proofs unchanged. He shows that ILP is in correspondence with the $\Box$-only fragment of the constructive logic CS4 as defined in [12].[1] Recently, Marti and Studer [26] supplied ILP with possible worlds semantics akin to the semantics developed by Fitting for the classical Logic of Proofs [17] and proved internalized disjunction property in its extension [27].

However, this axiomatization is not enough to obtain a proper intuitionistic arithmetical semantics, that is, to interpret '$t : A$' as 't is a proof of A in Heyting Arithmetic,' which is the motivation behind another line of work for considering intuitionistic versions of the Logic of Proofs. In order to obtain an intuitionistic Logic of Proofs complete for Heyting arithmetic, Artemov and Iemhoff [5] added to ILP extra axioms that internalize admissible rules of intuitionistic propositional logic. The arithmetical completeness was later shown by Dashkov [16]. Finally, Steren and Bonelli [31] provide an alternative system of terms for ILP based on natural deduction with hypothetical judgments.

What unifies all these versions of intuitionistic justification logics is the exclusive attention to the provability modality. Be the focus on semantics, realization theorem, or arithmetical completeness, the modal language is restricted to the $\Box$ modality. This restriction was quite natural in the classical setting, where $\Diamond$ can simply be viewed as the dual of $\Box$. However, with the freedom of De Morgan shackled comes the responsibility to treat $\Diamond$ as a fully independent modality—a responsibility that we take upon ourselves in this paper. In this first exploration of the kind of terms necessary to represent the operational side of the intuitionistic $\Diamond$ modality, we concentrate on *constructive versions* of several modal logics.[2]

Building on Artemov's treatment of the $\Box$-only fragment, we add a second type of terms, which we call *satisfier terms*, or simply *satisfiers*, and denote by Greek letters. Thus, a formula $\Diamond A$ is to be realized by '$\mu : A$.' The intuitive understanding of these terms is based on the view of $\Diamond$ modality as representing consistency (with $\Box$ still read as provability). A common way of proving consistency of a theory is to provide a model for this theory. Similarly, to prove that a formula is consistent with the theory, it is sufficient to present a model of the theory satisfying this formula. The satisfier μ

---

[1] Artemov himself called the logic CS4 "the intuitionistic modal logic on the basis of S4" and denoted it IS4.

[2] The reason for this is pragmatic: we discuss here only fragments which can be expressed in ordinary sequent calculus [35, 29, 12]. The more expressive intuitionistic modal logics require more elaborate sequent structures [32, 30]. We come back to this in the conclusion of this paper.

$$k_1: \; \Box(A \supset B) \supset (\Box A \supset \Box B)$$
$$k_2: \; \Box(A \supset B) \supset (\Diamond A \supset \Diamond B)$$

$$d: \; \Box A \supset \Diamond A$$
$$t: \; (A \supset \Diamond A) \wedge (\Box A \supset A)$$
$$4: \; (\Diamond\Diamond A \supset \Diamond A) \wedge (\Box A \supset \Box\Box A)$$

Figure 1: Modal axioms used in this paper

justifying the consistency of a formula is, therefore, viewed as an abstract model satisfying the formula. We keep these satisfying models abstract so as not to rely on any specific semantics. All the operations on satisfiers that we employ to ensure the realization theorem for CK, CD, CT, and CS4, as defined in [35, 29, 12], are akin to the operations on proof terms. In particular, the operation + for proof concatenation finds a counterpart in the operation $\sqcup$ for disjoint model union. Similarly, the application operation $\cdot$, which internalizes *modus ponens* reasoning by creating a new proof $t \cdot s$ for B from a given proof t of $A \supset B$ and a given proof s of A, has a counterpart $\star$ that creates a new *satisfier* $t \star \mu$ for B from a given proof t for $A \supset B$ and a given *satisfier* μ for A. The intuition behind this *satisfier propagation* operation $\star$ is that a proof of $A \supset B$, when applied to a satisfier for A provides evidence that the same model is also a satisfier for B. One could, perhaps, call it an internalized *model ponens*.

Outline of the paper: In Sect. 2, we introduce the syntax and proof theory of some constructive modal logics and, in Sect. 3, we give our definition of a justification logic for constructive modal logics. Then, in Sect. 4, we prove the main theorem of this paper, the realization theorem linking the various constructive modal logics to the corresponding justification logic. Finally, in Sect. 5, we point to further questions left as future work, as this paper is only the beginning of the research program consisting in giving justification logic for constructive and intuitionistic versions of modal logics.

2 Constructive modal logic

Let $a \in \mathcal{A}$ for a countable set of propositional variables $\mathcal{A}$. We define

$$A ::= \bot \mid a \mid (A \wedge A) \mid (A \vee A) \mid (A \supset A) \mid \Box A \mid \Diamond A$$

to be formulas in the modal language and use standard conventions regarding parentheses. We denote formulas by A, B, C, ... and define the negation as $\neg A := A \supset \bot$.

In modal logic, the behavior of the $\Box$ modality is determined by the k-axiom $\Box(A \supset B) \supset \Box A \supset \Box B$ and by the *necessitation rule* saying that, if A is valid, then so is $\Box A$, be the logic classical or intuitionistic. In classical modal logic the behavior of the $\Diamond$ modality is then fully determined by the De Morgan duality, which is violated in the intuitionistic case. This means that more axioms are needed to define the behavior of the $\Diamond$.

However, there is no unique way of doing so, and consequently many different variants of "intuitionistic modal logic" do exist. In this paper we consider the variant that is now called *constructive modal logic* [35, 12, 29, 2] and that is defined by adding to intuitionistic propositional logic the two axiom schemes shown in the left column of Fig. 1 together with the necessitation rule mentioned above. We call this logic CK. We also consider (i) the logic CD, which is CK extended with the d-axiom, (ii) the logic CT which is CK extended with the t-axiom, and (iii) the logic CS4 which is CT extended with the 4-axiom; all three axioms in the right column of Fig. 1.

Logics CK and CS4 are among those that have been studied most extensively. They can be given a possible world semantics by combining the interpretation of classical modal operators with that of intuitionistic implication. That is, a model for CK [28] is a tuple $(W, R, \leq, \models)$ where W is a set of worlds, R is a binary relation on W, $<$ is a preorder on W, and $\models$ is a relation between elements of W and formulas. In particular, in constructive modal logic, there can be *fallible* worlds in W such that $w \models \bot$. In a model for CS4 [1], R is additionaly reflexive and transitive (similarly to the case of classical S4) and the interaction of R and $\leq$ is constrained by the following relationship: $(R \circ \leq) \subseteq (\leq \circ R)$. To our knowledge, contrary to the classical case, the correspondence theory of CD and CT has not been investigated.

These logics have simple sequent calculi that can be obtained from any sequent calculus of intuitionistic propositional logic (IPL) by adding the appropriate rules for the modalities. In this paper, a *sequent* is an expression of the shape $B_1, \ldots, B_n \Rightarrow C$ where $B_1, \ldots, B_n$, and C are formulas and the antecedent to the left of $\Rightarrow$ has to be read as a multiset (i.e., the order of formulas is irrelevant, but it matters how often each formula appears). We use $\Gamma, \Delta, \Sigma, \ldots$ to denote such multisets of formulas. For a sequent $B_1, \ldots, B_n \Rightarrow C$ we define its *corresponding formula* $fm(B_1, \ldots, B_n \Rightarrow C)$ to be $B_1 \wedge \cdots \wedge B_n \supset C$. Most sequents in this paper consist of modal formulas. Thus, whenever we use the term "sequent" without any qualification, it is assumed that all formulas in it are modal formulas.

We start from the standard sequent calculus G3ip [33] whose rules are shown in Fig. 2. Then, the systems for the logics CK, CD, CT, and CS4, that we call LCK, LCD, LCT, and LCS4 respectively, are obtained by adding the rules in Fig. 3

$$\mathsf{id}\ \frac{}{\Gamma, a \Rightarrow a} \qquad\qquad \perp_\mathsf{L}\ \frac{}{\Gamma, \perp \Rightarrow C}$$

$$\vee_\mathsf{L}\ \frac{\Gamma, A \Rightarrow C \quad \Gamma, B \Rightarrow C}{\Gamma, A \vee B \Rightarrow C} \qquad \vee_\mathsf{R}\ \frac{\Gamma \Rightarrow A}{\Gamma \Rightarrow A \vee B} \quad \vee_\mathsf{R}\ \frac{\Gamma \Rightarrow B}{\Gamma \Rightarrow A \vee B}$$

$$\wedge_\mathsf{L}\ \frac{\Gamma, A, B \Rightarrow C}{\Gamma, A \wedge B \Rightarrow C} \qquad \wedge_\mathsf{R}\ \frac{\Gamma \Rightarrow A \quad \Gamma \Rightarrow B}{\Gamma \Rightarrow A \wedge B}$$

$$\supset_\mathsf{L}\ \frac{\Gamma, A \supset B \Rightarrow A \quad \Gamma, B \Rightarrow C}{\Gamma, A \supset B \Rightarrow C} \qquad \supset_\mathsf{R}\ \frac{\Gamma, A \Rightarrow B}{\Gamma \Rightarrow A \supset B}$$

Figure 2: Sequent calculus **G3ip** for intuitionistic propositional logic IPL

$$\mathsf{k}_\square\ \frac{\Gamma \Rightarrow A}{\square\Gamma, \Delta \Rightarrow \square A} \qquad \mathsf{k}_\diamond\ \frac{\Gamma, B \Rightarrow A}{\square\Gamma, \Delta, \diamond B \Rightarrow \diamond A} \qquad \mathsf{d}\ \frac{\Gamma \Rightarrow A}{\square\Gamma, \Delta \Rightarrow \diamond A}$$

$$\mathsf{4}_\square\ \frac{\square\Gamma \Rightarrow A}{\square\Gamma, \Delta \Rightarrow \square A} \quad \mathsf{4}_\diamond\ \frac{\square\Gamma, B \Rightarrow \diamond A}{\square\Gamma, \Delta, \diamond B \Rightarrow \diamond A} \quad \mathsf{t}_\square\ \frac{\Gamma, \square A, A \Rightarrow B}{\Gamma, \square A \Rightarrow B} \quad \mathsf{t}_\diamond\ \frac{\Gamma \Rightarrow A}{\Gamma \Rightarrow \diamond A}$$

Figure 3: Additional rules for modalities

according to the following table.[3]

$$
\begin{aligned}
\mathsf{LCK} &= \mathsf{G3ip} + \mathsf{k}_\square + \mathsf{k}_\diamond \\
\mathsf{LCD} &= \mathsf{G3ip} + \mathsf{k}_\square + \mathsf{k}_\diamond + \mathsf{d} \\
\mathsf{LCT} &= \mathsf{G3ip} + \mathsf{k}_\square + \mathsf{k}_\diamond + \mathsf{t}_\square + \mathsf{t}_\diamond \\
\mathsf{LCS4} &= \mathsf{G3ip} + \mathsf{4}_\square + \mathsf{4}_\diamond + \mathsf{t}_\square + \mathsf{t}_\diamond
\end{aligned}
\tag{1}
$$

Observe that the axiom rule id is restricted to atomic formulas. We rely on that in the proof of the realization theorem in Sect. 4. However, as expected, using the standard argument by induction on the formula construction, the general form of the axiom rule is derivable.

Lemma 2.1 (Generalized axioms). *For every formula A, the rule* $\mathsf{id}_g\ \frac{}{\Gamma, A \Rightarrow A}$ *is derivable in each of* **G3ip**, **LCK**, **LCD**, **LCT**, *and* **LCS4**.

[3]For a survey of the classical variants of these systems, see, for example, [34].

Finally, the rule   cut $\dfrac{\Gamma \Rightarrow A \qquad \Delta, A \Rightarrow C}{\Gamma, \Delta \Rightarrow C}$ is admissible.

Theorem 2.2 (Cut Admissibility). *Let* LML $\in$ {LCK, LCD, LCT, LCS4}. *If a sequent is provable in* LML + cut *then it is also provable in* LML.

Proof. For LCK, LCD, and LCT, the proof follows as a special case of [25], and for CS4 the result is stated in [12] as a "routine adaptation of Gentzen's method." $\qquad\square$

Using Theorem 2.2, we can easily show the completeness of our system.

Theorem 2.3 (Completeness). *Let* ML $\in$ {CK, CD, CT, CS4} *and* LML *be the corresponding sequent system. If* $\vdash_{ML} A$, *then* $\vdash_{LML} \Rightarrow A$.

Proof. The axioms of IPL can be proved using G3ip in Fig. 2; those in Fig. 1 can be proved using the corresponding rules in Fig. 3. Finally, the necessitation rule can be simulated with $k_\square$, and *modus ponens* can be simulated using cut. Now completeness of the cut-free systems follows immediately from Theorem 2.2. $\qquad\square$

Theorem 2.4 (Soundness). *Let* ML $\in$ {CK, CD, CT, CS4}. *If* $B_1, \ldots, B_n \Rightarrow C$ *is a sequent provable in the corresponding sequent system* LML, *then* $B_1 \wedge \cdots \wedge B_n \supset C$ *is a theorem of* ML.

Proof. We proceed by induction on the proof π of $B_1, \ldots, B_n \Rightarrow C$ in LML, making a case analysis on the bottom-most rule instance in π. For the rules in G3ip, this is straightforward. Now consider the rule $\quad k_\square \dfrac{C_1, \ldots, C_n \Rightarrow A}{\square C_1, \ldots, \square C_n, D_1, \ldots, D_m \Rightarrow \square A}$.
By induction hypothesis, $\vdash_{ML} C_1 \wedge \cdots \wedge C_n \supset A$, hence by intuitionistic reasoning, $\vdash_{ML} C_1 \supset \cdots \supset C_n \supset A$.[4] By necessitation, $\vdash_{ML} \square(C_1 \supset \cdots \supset C_n \supset A)$, and, using k_1 and *modus ponens*, we get $\vdash_{ML} \square C_1 \supset \cdots \supset \square C_n \supset \square A$. Hence, $\vdash_{ML} \square C_1 \wedge \cdots \wedge \square C_n \wedge D_1 \wedge \cdots \wedge D_m \supset \square A$ follows by intuitionistic reasoning. Other cases are similar. $\qquad\square$

3   Justification logic

Justification logic adds *proof terms* directly inside its language using formulas '$t : A$' with the meaning 't *is a proof of* A.' In the constructive version that we propose in this section, we will also add *satisfiers* into the language, using formulas '$\mu : A$' with the underlying intuition that 'μ *is a model of* A.'

---

[4]Throughout the paper we consider $\supset$ to be right-associative.

Proof terms, intended to replace $\Box$, are denoted $t, s, \ldots$, while *satisfiers*, intended to replace $\Diamond$, are denoted $\mu, \nu, \ldots$ Proof terms are built from a set of *proof variables*, denoted $x, y, \ldots$, and a set of *(proof) constants*, denoted $c, d, \ldots$, using the operations *application* $\cdot$, *sum* $+$, and *proof checker* !. Satisfiers are built from a set of *satisfier variables*, denoted $\alpha, \beta, \ldots$, using the operations *disjoint union* $\sqcup$ (binary operation on satisfiers) and *propagation* $\star$ (combines a proof term with a satisfier).

$$
\begin{aligned}
t &::= \quad c \quad | \quad x \quad | \quad (t \cdot t) \quad | \quad (t + t) \quad | \quad !t \\
\mu &::= \qquad\quad \alpha \quad | \quad (t \star \mu) \quad | \quad (\mu \sqcup \mu)
\end{aligned}
$$

While the intuitive meaning of the operations $\cdot$, $+$, and ! on proof terms has been well documented in justification logic literature and corresponds to rather well known proof manipulations, it is worth explaining our intuition behind the new operations $\star$ and $\sqcup$ involving satisfiers.

The operation $\star$ is a combination of global and local reasoning. For instance, assume that $\neg\neg A$ is true; therefore, by classical propositional logic, A must be true. Here $\neg\neg A$ being true is a local, contingent fact, whereas the transition is made based on the classical tautology $\neg\neg A \supset A$. The result is the contingent truth of A in the same situation where $\neg\neg A$ is true. We are working in a language with explicit proofs for valid statements and explicit satisfiers representing specific models satisfying a statement. Thus, given a satisfier μ for A and a proof t that generally $A \supset B$, we can conclude B. While B is true whenever A is, the justification used is different in that the former involves a valid transition from A to B justified by t. Hence, instead of using the same satisfier μ, we record our reasoning in the new satisfier $t \star \mu$. For instance, if satisfiers are in principle intended to range over intuitionistic Kripke models, then $x : (\neg\neg A \supset A)$ becomes a non-trivial assumption on whether only classical models are considered. Hence, the truth of A depends not only on the truth of $\neg\neg A$ in a model represented by the satisfier μ but also on the validity of the law of double negation.

The operation $\sqcup$ of disjoint model union is akin to that of disjoint set union. For instance, for sets, one often defines $X \sqcup Y = (X \times \{0\}) \cup (Y \times \{1\})$ in order to avoid potential problems of X overlapping with Y and be able to state facts such as $|X \sqcup Y| = |X| + |Y|$. Intuitively, our disjoint model union works the same way. Whatever the nature of models represented by satisfiers μ and ν, any overlaps among them are resolved before the models are combined and no connection between the μ and ν parts of the satisfier $\mu \sqcup \nu$ exists. For instance, the disjoint union of intuitionistic Kripke models $\mathcal{M}_1 = (W_1, \leq_1, V_1)$ and $\mathcal{M}_2 = (W_2, \leq_2, V_2)$ can be defined as follows: $\mathcal{M}_1 \sqcup \mathcal{M}_2 := (W, \leq, V)$ where $W := (W_1 \times \{0\}) \cup (W_2 \times \{1\})$, $(w, i) \leq (w', j)$ iff $i = j$ and $w \leq_i w'$, and $V((w, i)) := V_i(w)$.

taut:	Complete finite set of axioms for IPL
$jk_\square$:	$t:(A \supset B) \supset (s:A \supset t \cdot s:B)$
$jk_\diamond$:	$t:(A \supset B) \supset (\mu:A \supset t \star \mu:B)$
sum:	$s:A \supset (s+t):A$ and $t:A \supset (s+t):A$
union:	$\mu:A \supset (\mu \sqcup \nu):A$ and $\nu:A \supset (\mu \sqcup \nu):A$

$$\mathsf{mp}\ \frac{A \supset B \quad B}{B} \qquad \mathsf{ian}\ \frac{A \text{ is an axiom instance}}{c_1:\ldots c_n:A}$$

Figure 4: Axiomatization of the constructive justification logic JCK

$jd_\square$:	$t:\bot \supset \bot$	$jd_\diamond$:	$\top \supset \mu:\top$
$jt_\square$:	$t:A \supset A$	$jt_\diamond$:	$A \supset \mu:A$
$j4_\square$:	$t:A \supset\ !t:t:A$	$j4_\diamond$:	$\mu:\nu:A \supset \nu:A$

Figure 5: Additional justification axioms

The formulas of justification logic are obtained from the following grammar:

$$A ::= \bot \mid a \mid (A \wedge A) \mid (A \vee A) \mid (A \supset A) \mid t:A \mid \mu:A$$

We propose to extend the formulation of justification logics to realize constructive modal logics. The axiomatization of the basic one is shown in Fig. 4. It is similar to the standard justification counterpart of the classical modal logic K except for the additional axiom $jk_\diamond$, which corresponds to the modal axiom k_2. The other axioms taut, $jk_\square$, and sum, as well as the rules of *modus ponens* mp and *iterated axiom necessitation* ian are standard, e.g., see [22]. We call this basic logic JCK, and as in the classical setting, we can define extension of JCK using the axioms defined in Fig. 5. The logic JCD is obtained from JCK by adding the axioms $jd_\square$ and $jd_\diamond$; the logic JCT is obtained from JCK by adding the axioms $jt_\square$ and $jt_\diamond$; and the logic JCS4 is obtained by adding the axioms $j4_\square$ and $j4_\diamond$ to JCT. Note that the $\square$ variant of each axiom corresponds exactly to the one used in the classical setting. Our contribution is the definition of the $\diamond$ variants operating on the satisfiers instead of the proof terms.

The intuitive reading of these new satisfier axioms is as follows. The axiom $jd_\diamond$ states that $\top$ is satisfied in every model. The axiom $jt_\diamond$ could be understood as the insistence that the actual model must be part of any other model considered: if A is true, then it is satisfied in every model. Perhaps, the least intuitive is the axiom $j4_\diamond$. One way of reading it is to say that truth in models is "context-free." The fact of A being satisfied in a model represented by ν does not depend on ν being considered within the context of another model represented by μ. Put another way, any sub-model ν of μ can also be considered in isolation and produces the same truth values.

The logics JCK, JCD, JCT, and JCS4 can be seen as the operational version of the constructive modal logics CK, CD, CT, and CS4 respectively, defined in the previous section. Indeed if one forgets about the proof term and satisfier annotations and considers them as empty $\Box$ and $\Diamond$ respectively, the logics prove the same theorems.

Definition 3.1. We define the operation of *forgetful projection* $(\cdot)^\circ$ that maps justification formulas onto corresponding modal formulas recursively: $\bot^\circ := \bot$, $a^\circ := a$ for all propositional variables a, $(t : A)^\circ := \Box A^\circ$, $(\mu : A)^\circ := \Diamond A^\circ$, and for $* \in \{\wedge, \vee, \supset\}$, finally, $(A * B)^\circ := A^\circ * B^\circ$.

We extend this definition to multisets of formulas: $(A_1, \ldots, A_n)^\circ := A_1^\circ, \ldots, A_n^\circ$.

It is easy to show by induction on the Hilbert derivation in JL that

Lemma 3.2 (Forgetful projection). *Let* JL $\in \{$JCK, JCD, JCT, JCS4$\}$ *and* ML *be the corresponding modal logic. If* $\vdash_{\mathsf{JL}} F$, *then* $\vdash_{\mathsf{ML}} F^\circ$.

The more difficult question however is: can we prove the converse? This result is called realization, namely that every theorem of a certain modal logic can be 'realized' by a justification theorem. However, it is not such an easy result as it may seem. It is not possible to directly transform a Hilbert proof of a modal theorem into a Hilbert proof of its realization in justification logic as the rule mp in a Hilbert system can create dependencies between modalities. The standard solution to this issue is to consider a proof of the modal theorem in a *cut-free* sequent calculus as the absence of cuts in the proof will prevent the creation of dependencies. The detailed statement and proof of this result can only be presented in the next section, as we have to introduce some basics first.

We state below two lemmas that are crucial for the realization proof: the *Lifting Lemma* and the *Substitution Property*. They are extensions of standard results from the justification logics literature to the constructive case. Repeating verbatim the proof from [7], we obtain the *Lifting Lemma* and its variant showing that necessitation can be internalized within the language of these justification logics.

Lemma 3.3 (Lifting Lemma). *Let* JL $\in \{$JCK, JCD, JCT, JCS4$\}$. *If*

$$A_1, \ldots, A_n \vdash_{\mathsf{JL}} B,$$

then there exists a proof term $t(x_1, \ldots, x_n)$ *such that for all proof terms* $s_1, \ldots, s_n$

$$s_1 : A_1, \ldots, s_n : A_n \vdash_{\mathsf{JL}} t(s_1, \ldots, s_n) : B.$$

Corollary 3.4. *Let* $\mathsf{JL} \in \{\mathsf{JCK}, \mathsf{JCD}, \mathsf{JCT}, \mathsf{JCS4}\}$. *If* $\vdash_{\mathsf{JL}} A_1 \wedge \cdots \wedge A_n \supset B$, *then there exists a proof term* $t(x_1, \ldots, x_n)$ *such that for all proof terms* $s_1, \ldots, s_n$ *we have* $\vdash_{\mathsf{JL}} s_1 : A_1 \wedge \cdots \wedge s_n : A_n \supset t(s_1, \ldots, s_n) : B$.

In our constructive setting, we also need a $\diamond$ variant of this statement.

Corollary 3.5. *Let* $\mathsf{JL} \in \{\mathsf{JCK}, \mathsf{JCD}, \mathsf{JCT}, \mathsf{JCS4}\}$. *If*

$$\vdash_{\mathsf{JL}} A_1 \wedge \cdots \wedge A_n \wedge C \supset B,$$

then there is a satisfier $\mu(x_1, \ldots, x_n, \beta)$ *such that for all proof terms* $s_1, \ldots, s_n$ *and any satisfier* ν

$$\vdash_{\mathsf{JL}} s_1 : A_1 \wedge \cdots \wedge s_n : A_n \wedge \nu : C \supset \mu(s_1, \ldots, s_n, \nu) : B. \tag{2}$$

Proof. By intuitionistic reasoning and Cor. 3.4, we get a proof term $t(x_1, \ldots, x_n)$ such that

$$\vdash_{\mathsf{JL}} s_1 : A_1 \wedge \cdots \wedge s_n : A_n \supset t(s_1, \ldots, s_n) : (C \supset B).$$

Using the instance $t(s_1, \ldots, s_n) : (C \supset B) \supset \nu : C \supset (t(s_1, \ldots, s_n) \star \nu) : B$ of the axiom $\mathsf{jk}_\diamond$, we can see that (2) holds for $\mu(x_1, \ldots, x_n, \beta) := t(x_1, \ldots, x_n) \star \beta$. $\square$

Finally, we generalize the standard definition of substitution to our setting.

Definition 3.6. A *substitution* σ maps proof variables to proof terms and satisfier variables to satisfiers. The application of a substitution σ to a proof term t or satisfier μ, denoted $t\sigma$ or $\mu\sigma$ respectively, is defined recursively as follows:

$$
\begin{aligned}
c\sigma &:= c & x\sigma &:= \sigma(x) \\
(t \cdot s)\sigma &:= t\sigma \cdot s\sigma & (t + s)\sigma &:= t\sigma + s\sigma \\
(!t)\sigma &:= !(t\sigma) & \alpha\sigma &:= \sigma(\alpha) \\
(t \star \mu)\sigma &:= t\sigma \star \mu\sigma & (\mu \sqcup \nu)\sigma &:= \mu\sigma \sqcup \nu\sigma
\end{aligned}
$$

where c is a proof constant, x is a proof variable, and α is a satisfier variable. The application of σ to a justification formula A yields the formula $A\sigma$, where each proof term t (respectively satisfier μ) appearing in A is replaced with $t\sigma$ (respectively $\mu\sigma$).

The proof of the Substitution Property from [7] is easily adaptable to our case:

Lemma 3.7 (Substitution Property). *Let* $\mathsf{JL} \in \{\mathsf{JCK}, \mathsf{JCD}, \mathsf{JCT}, \mathsf{JCS4}\}$. *If* $\vdash_{\mathsf{JL}} A$, *then* $\vdash_{\mathsf{JL}} A\sigma$ *for any substitution* σ.

Remark 3.8. In our formulation, the Substitution Property holds because the rule ian is formulated in its strongest form, with all proof constants being interchangeable. Combined with the schematic formulation of all axioms, this makes derivations impervious to substitutions. A more nuanced formulation would be to restrict ian to a specific set of instances collected in a *constant specification* (our variant corresponds to the *total constant specification*). It is a standard fact in justification logic that the substitution property only holds for *schematic* constant specifications, i.e., those invariant with respect to substitutions. The only difference for our logics is that a schematic constant specification must additionally be schematic with respect to substitutions of satisfiers for satisfier variables.

4 Realization theorem for constructive modal logic

Assume we have a justification formula F and its forgetful projection F°. In that case we call F a *realization* of F°. Similarly, a *justification sequent* $\Gamma \Rightarrow C$, that is, a sequent consisting of justification formulas, can be the *realization* of a modal sequent $\Gamma^\circ \Rightarrow C^\circ$. In order to define the notion of *normal realization* we need the notions of positive and negative occurrences of subformulas.

An occurrence of a subformula A of F is called *positive* if the position of A in the syntactic tree of F is reached from the root by following the left branch of an $\supset$ branching an even number of times; otherwise it is called *negative*. For example, the displayed subformula A is positive in the formula $(A \supset B) \supset C$ but negative in the formula $A \supset (B \supset C)$. The *polarity* of the occurrence of a subformula in a sequent $\Gamma \Rightarrow C$ is given by its polarity in the formula $fm(\Gamma \Rightarrow C)$.

Definition 4.1. A realization $\Gamma \Rightarrow C$ of $\Gamma^\circ \Rightarrow C^\circ$ is called *normal* if the following condition is fulfilled: if $t : A$ (respectively $\mu : A$) is a negative subformula occurrence of $\Gamma \Rightarrow C$, then t is a proof variable (respectively μ is a satisfier variable) that occurs in $\Gamma \Rightarrow C$ exactly once.

We can now state and prove the main theorem of this paper.

Theorem 4.2 (Realization). *Let* ML $\in \{$CK, CD, CT, CS4$\}$, JL *be the corresponding justification logic, i.e.,* JCK, JCD, JCT, *or* JCS4 *respectively, and* LML *be the cut-free sequent calculus for* ML. *If* $\vdash_{\mathsf{LML}} \Gamma' \Rightarrow C'$ *for a given modal sequents, then there is a normal realization* $\Gamma \Rightarrow C$ *of* $\Gamma' \Rightarrow C'$ *such that* $\vdash_{\mathsf{JL}} fm(\Gamma \Rightarrow C)$.

Corollary 4.3. *Let* ML $\in \{$CK, CD, CT, CS4$\}$ *and* JL *be the corresponding justification logic. If* $\vdash_{\mathsf{ML}} A$, *then* $\vdash_{\mathsf{JL}} F$ *for some justification formula* F *such that* $F^\circ = A$.

Proof of Theorem 4.2. The proof goes largely along the lines of that for the $\Box$-only classical fragment (see [7, 13]). The operation $\sqcup$ on satisfiers plays the same role as the operation $+$ on proof terms. Thus, we only show in detail cases for the new rules. As a matter of a shorthand, we say that a justification sequent $\Gamma \Rightarrow C$ is *derivable in* JL if its corresponding formula is, i.e., if $\vdash_{\mathsf{JL}} fm(\Gamma \Rightarrow C)$.

Let π be an LML proof of $\Gamma' \Rightarrow C'$. We assign a unique index $i \in \{1, \ldots, n\}$ to each of n occurrences of $\Box$ and $\Diamond$ in its endsequent $\Gamma' \Rightarrow C'$. We define the *modal flow graph* of π, denoted G_π, as follows: its vertices are all occurrences of formulas of the form $\Box A$ and $\Diamond A$ in π. Two such occurrences are connected with an edge iff they are occurrences of the same formula within the same rule instance and

- either one occurs within a side formula in a premise and the other is the same occurrence within the same subformula in the conclusion

- or one occurs within an active formula in a premise and the other is the corresponding occurrence within the principal formula in the conclusion.

Each connected component of G_π has exactly one vertex in the endsequent of π and all vertices in the connected component are assigned the same index as this representative in the endsequent. E.g., in the following instance of $\mathsf{k}_\Box$, modalities connected by edges are vertically aligned and given the same index:

$$\mathsf{k}_\Box \frac{\Box_5 a \lor \Diamond_7 b \, , \qquad c \supset d \qquad\qquad\qquad \Rightarrow \qquad \Diamond_9 g \supset \Box_8 h}{\Box_2(\Box_5 a \lor \Diamond_7 b), \Box_6(c \supset d), \Box_3 e, \Box_{10}\Diamond_{20}f \Rightarrow \Box_{15}(\Diamond_9 g \supset \Box_8 h)} \qquad (3)$$

In the absence of the cut rule, the resulting graph is a forest where each tree has its root in the endsequent and is identified with a unique modality type $\heartsuit$ and unique index i. We denote it a $\heartsuit_i$-*tree*. Branching occurs in the branching rules, as well as in the rules with embedded contraction, e.g., in $\mathsf{t}_\Box$ each modality in A within $\Box A$ in the conclusion of the rule branches to the corresponding occurrence in A and the corresponding occurrence in $\Box A$ in the premise. Each leaf of a $\heartsuit_i$-tree is either in a side formula of an axiom id or $\bot_\mathsf{L}$, in which case it is called an *initial leaf*, or in the conclusion of a modal rule from Fig. 3 that introduced $\heartsuit_i$, in which case it is called a *modal leaf*. For instance, if (3) is used in π, then the $\Box_2$-, $\Box_6$-, $\Box_3$-, $\Box_{10}$-, $\Diamond_{20}$-, and $\Box_{15}$-trees in G_π have modal leaves in the conclusion of (3).

We call the number of modal leaves of a $\heartsuit_i$-tree occurring in the succedents of modal rules the *multiplicity of* i, denoted by m_i, which is a non-negative integer.

From the tree π of modal sequents, we construct another tree π_0 of justification sequents by replacing

each $\Box_i$ for $m_i > 0$ with $z_i := y_{i,1} + \cdots + y_{i,m_i}$ for proof variables $y_{i,1}, \ldots, y_{i,m_i}$;

each $\Box_i$ for $m_i = 0$ with $z_i := y_{i,0}$ for a proof variable $y_{i,0}$;

each $\Diamond_i$ for $m_i > 0$ with $\omega_i := \beta_{i,1} \sqcup \cdots \sqcup \beta_{i,m_i}$ for satisfier variables $\beta_{i,1}, \ldots, \beta_{i,m_i}$;

each $\Diamond_i$ for $m_i = 0$ with $\omega_i := \beta_{i,0}$ for a satisfier variable $\beta_{i,0}$.

All proof variables $y_{i,j}$ and all satisfier variables $\beta_{i,j}$ must be pairwise distinct.

Let us call a rule *justificational* if it is one of $k_\Box$, $k_\Diamond$, d, or $4_\Box$. All other rules, including the rules $4_\Diamond$, $t_\Box$, and $t_\Diamond$, as well as the rules in Fig. 2 are *simple*. Let k be the number of instances of justificational rules in π. We will construct a sequence of substitutions $\sigma_1, \ldots, \sigma_k$ that, when applied to π_0, produces a sequence $\pi_1, \ldots, \pi_k$ of trees such that $\pi_{h+1} = \pi_h \sigma_{h+1}$. Note that for any justification sequent in the tree π_h, its forgetful projection is the modal sequent from the corresponding node of the tree π and that every occurrence of $\Box_i$ or $\Diamond_i$ in π is replaced in π_h with $z_i \sigma_1 \ldots \sigma_h$ or $\omega_i \sigma_1 \ldots \sigma_h$ respectfully. For $\tau_h := \sigma_h \circ \cdots \circ \sigma_1$ let us call $z_i \tau_h$ and $\omega_i \tau_h$ the *h-prerealizations* of $\Box_i$, and $\Diamond_i$ respectively.[5] For any sequent occurrence $\Delta \Rightarrow D$ in π, we call $(\Delta \Rightarrow D)\tau_h$ its h-prerealization and denote it $\Delta_h \Rightarrow D_h$.

Let the k justificational rules be ordered linearly in a way consistent with the tree order of π: for arbitrary $k \geq j > i \geq 1$, the jth rule is not inside a subtree rooted at the premise of the ith rule. By induction on $i = 0, \ldots, k$ we will show that,

1. for any subtree of π_i with no occurrences of modal rules $i+1, \ldots, k$ the end-sequent $\Delta_i \Rightarrow D_i$ of this subtree is derivable in JL, i.e., h-prerealizations of a sequent occurrence $\Delta \Rightarrow D$ from π become derivable as soon as h overtakes the numbers of all justificational rules used to derive the sequent in π;

2. $y_{i,0}\tau_h = y_{i,0}$ and $\beta_{i,0}\tau_h = \beta_{i,0}$ for all justificational rules above $h = 1, \ldots, k$, i.e., terms prerealizing modalities not contributing to m_i remain fixed points for all substitutions.

In particular, after all justificational rules are processed in π_k, the k-prerealization $\Gamma_k \Rightarrow C_k$ of the endsequent of π will be derivable in JL making it a realization. Moreover, since no negative occurrence of a modality from the endsequent can be traced to a leaf in a succedent of a sequent from π, in this realization all such negative modalities are realized by proof and satisfier variables. We prove it by a secondary induction on the depth of the proof up to the first unprocessed justificational rule.

For a simple rule, the JL-derivability of the i-prerealization of its premise(s) implies the JL-derivability of the i-prerealization of its conclusion. For the rules from Fig. 2 the reasoning is propositional. For rules $t_\Box$ and $t_\Diamond$ (applicable only to LCT

---

[5]The term prerealization is used here in its layman's meaning of an almost but not quite a realization and is unrelated to the use in [22].

and LCS4), it follows by axioms $jt_\Box$ and $jt_\Diamond$ of JCT and JCS4. For LCS4, assume that for a rule instance $4_\Diamond \dfrac{\Box_{k_1} G^1, \ldots, \Box_{k_r} G^r, B \Rightarrow \Diamond_j A}{\Box_{k_1} G^1, \ldots, \Box_{k_r} G^r, D^1, \ldots, D^p, \Diamond_l B \Rightarrow \Diamond_j A}$ from π the i-prerealization of the premise is derivable in JL, i.e.,

$$\vdash_{\mathsf{JL}} z_{k_1} : G_i^1 \wedge \cdots \wedge z_{k_r} : G_i^r \wedge B_i \supset \omega_j \tau_i : A_i.$$

By Cor. 3.5, there is a satisfier $\mu(x_1, \ldots, x_r, \beta)$ such that

$$\vdash_{\mathsf{JL}} \,! z_{k_1} : z_{k_1} : G_i^1 \wedge \cdots \wedge \,! z_{k_r} : z_{k_r} : G_i^r \wedge \omega_l : B_i \supset \mu(! z_{k_1}, \ldots, ! z_{k_r}, \omega_l) : \omega_j \tau_i : A_i \ .$$

It now follows by $j4_\Box$ and $j4_\Diamond$ of JCS4 and propositional reasoning that

$$\vdash_{\mathsf{JL}} z_{k_1} : G_i^1 \wedge \cdots \wedge z_{k_r} : G_i^r \wedge D_i^1 \wedge \cdots \wedge D_i^p \wedge \omega_l : B_i \supset \omega_j \tau_i : A_i$$

making the i-prerealization of the conclusion of the rule derivable in JL.

This observation alone establishes the base of the main induction, i.e., that all 0-prerealizations of modal sequents derived without the use of justificational rules are derivable in JL.

For the step of the main induction, consider the premise of the hth justificational rule and assume its $(h-1)$-prerealization is derivable by IH. For each of the justificational rules we will show how to apply an additional substitution to make its conclusion derivable. By the Substitution Property (Lemma 3.7), this substitution preserves the derivability of all h-prerealizations of modal sequents whose $(h-1)$-prerealizations are derivable by the IH, including the premise of the hth justificational rule. Thus, the h-prerealization of its conclusion is also derivable and the argument about simple rules can be applied to extend this result down until the next justificational rule. The cases of the $k_\Box$ and $4_\Box$ rules are treated the same way as in [13] by means of Cor. 3.4. It remains to process the two remaining justificational rules.

We start with the case where the hth rule in π is the qth introduction of $\Diamond_j$ by a justificational rule out of m_j: $k_\Diamond \dfrac{G^1, \ldots, G^r, B \Rightarrow A}{\Box_{k_1} G^1, \ldots, \Box_{k_r} G^r, D^1, \ldots, D^p, \Diamond_l B \Rightarrow \Diamond_j A}.$
Assume that the $(h-1)$-prerealization of the premise is derivable in JL, i.e.,

$$\vdash_{\mathsf{JL}} G_{h-1}^1 \wedge \cdots \wedge G_{h-1}^r \wedge B_{h-1} \supset A_{h-1} \ . \tag{4}$$

By Cor. 3.5 there is a satisfier $\mu(x_1, \ldots, x_r, \beta)$ such that

$$\vdash_{\mathsf{JL}} z_{k_1} : G_{h-1}^1 \wedge \cdots \wedge z_{k_r} : G_{h-1}^r \wedge \omega_l : B_{h-1} \supset \mu(z_{k_1}, \ldots, z_{k_r}, \omega_l) : A_{h-1}.$$

We define $\sigma_h \colon \beta_{j,q} \mapsto \mu(z_{k_1}, \ldots, z_{k_r}, \omega_l)$. Note that σ_h affects exactly one satisfier variable, which is neither $y_{i,0}$ nor $\beta_{i,0}$ and which corresponds to the justificational rule being processed. In particular, $\beta_{j,q}\tau_{h-1} = \beta_{j,q}$ and $\beta_{j,q}\tau_h = \beta_{j,q}\sigma_h$. Thus,

$$\vdash_{\mathsf{JL}} z_{k_1} : G^1_{h-1} \wedge \cdots \wedge z_{k_r} : G^r_{h-1} \wedge \omega_l : B_{h-1} \supset \beta_{j,q}\tau_h : A_{h-1}.$$

Applying σ_h substitution, we obtain by the Substitution Property,

$$\vdash_{\mathsf{JL}} z_{k_1} : (G^1_{h-1}\sigma_h) \wedge \cdots \wedge z_{k_r} : (G^r_{h-1}\sigma_h) \wedge \omega_l : (B_{h-1}\sigma_h) \supset \beta_{j,q}\tau_h : (A_{h-1}\sigma_h)$$

because (a) σ_h does not affect the proof variables $z_{k_1}, \ldots, z_{k_r}$, (b) σ_h does not affect the satisfier variable $\omega_l \neq \beta_{j,q}$ because j and l are indices of diamonds of opposite polarity, and (c) σ_h does not affect the satisfier $\beta_{j,q}\tau_h = \mu(z_{k_1}, \ldots, z_{k_r}, \omega_l)$ because the only variables occurring in it are $z_{k_1}, \ldots, z_{k_r}$, and ω_l. It follows by **union** that

$$\vdash_{\mathsf{JL}} z_{k_1} : G^1_h \wedge \cdots \wedge z_{k_r} : G^r_h \wedge D^1_h \wedge \cdots \wedge D^p_h \wedge \omega_l : B_h \supset \omega_j\tau_h : A_h$$

where $\omega_j = \beta_{j,1} \sqcup \cdots \sqcup \beta_{j,q} \sqcup \cdots \sqcup \beta_{j,m_j}$. Thus, the h-realization of the conclusion is also derivable in JL.

The case of the rule $\mathsf{d} \dfrac{G^1, \ldots, G^r \Rightarrow A}{\square_{k_1} G^1, \ldots, \square_{k_r} G^r, D^1, \ldots, D^p, \Rightarrow \lozenge_j A}$ for LCD is similar. By the IH, (4) holds for $B_{h-1} = \top$. Repeating all the steps for $\mathsf{k}_\lozenge$ and using a fresh satisfier variable β in place of ω_l for $\lozenge\top$, we obtain

$$\vdash_{\mathsf{JL}} z_{k_1} : G^1_h \wedge \cdots \wedge z_{k_r} : G^r_h \wedge D^1_h \wedge \cdots \wedge D^p_h \wedge \beta : \top \supset \omega_j\tau_h : A_h \ .$$

It remains to note that $\vdash_{\mathsf{JL}} \beta : \top$ by axiom $\mathsf{jd}_\lozenge$ of JCD. It follows that

$$\vdash_{\mathsf{JL}} z_{k_1} : G^1_h \wedge \cdots \wedge z_{k_r} : G^r_h \wedge D^1_h \wedge \cdots \wedge D^p_h \supset \omega_j\tau_h : A_h \ . \qquad \square$$

The crucial difference between justificational and simple rules is that, unlike the former, the latter require an additional substitution on top of all the previous ones.

5 Conclusion and future work

In this paper, we proposed justification counterparts for some constructive modal logics, which, for the first time, employ the notion of satisfiers to realize the $\lozenge$-modality. This led us to define an operator combining proof terms and satisfiers, which is crucial to the realization of the constructive modal axiom k_2. However, surprisingly, the only other operation needed on satisfiers is the disjoint union, an equivalent to the sum for proof terms. In particular, while the $\square$-version of the 4-axiom traditionally requires the proof checker operator !, the $\lozenge$-version of axiom 4 does not seem to necessitate any additional operation on satisfiers. In the following, we list a handful of directions for future work:

$$k4_\square \frac{\square\Gamma, \Gamma \Rightarrow A}{\Delta, \square\Gamma \Rightarrow \square A} \qquad k4_\lozenge \frac{\square\Gamma, \Gamma, B \Rightarrow A}{\Delta, \square\Gamma, \lozenge B \Rightarrow \lozenge A} \qquad k4'_\lozenge \frac{\square\Gamma, \Gamma, B \Rightarrow \lozenge A}{\Delta, \square\Gamma, \lozenge B \Rightarrow \lozenge A}$$

Figure 6: More rules for modalities

- Semantics of our proposed logics. Modular models from [10, 23] should provide a good starting point, but require significant adjustments.

- We have chosen to work with the logics that have simple, known cut-free sequent calculi, a property on which the realization proof strongly relies. The same method can be further extended to CK4 and CD4 that are obtained from CK and CD, respectively, by adding the 4-axiom. To our knowledge, these logics have not been independently studied, but it is possible to 'constructivize' the classical rule $k4_\square$ in the same way as for the rules in Fig. 3. That is, corresponding sequent systems to CK4 and CD4 may be obtained via the rules in Fig. 6:

$$
\begin{aligned}
\text{LCK4} &= \text{G3ip} + k4_\square + k4_\lozenge + k4'_\lozenge \\
\text{LCD4} &= \text{G3ip} + k4_\square + k4_\lozenge + k4'_\lozenge + d
\end{aligned}
$$

We decided to forgo this extension for pragmatic reasons: without a cut-free calculi for these constructive modal logics in the literature we would need to provide a full cut-elimination proof. Even though it should be possible to directly adapt for example the proof from [25], it would have changed the focus of this paper.

- There exist other, more elaborate realization proofs, e.g., from [19], that provide realizations with additional properties and/or structure. Applying them to modal logics with non-classical propositional basis remains future work.

- We believe that our way of justifying the $\lozenge$ modality would similarly work for the "intuitionistic variant" of modal logic [30], which is obtained from the constructive variant by adding the three axioms $k_3 : \lozenge(A \vee B) \supset (\lozenge A \vee \lozenge B)$ and $k_4 : (\lozenge A \supset \square B) \supset \square(A \supset B)$ and $k_5 : \lozenge\bot \supset \bot$. There are no ordinary sequent calculi for such logics, so the proof of realization provided here could not be straightforwardly adapted. However, there are nested sequent calculi for all logics in the *intuitionistic S5-cube* [32], even in a focused variant [15], which means that we might still be able to prove a realization theorem by extending the method used in [22].

References

[1] Natasha Alechina, Michael Mendler, Valeria de Paiva, and Eike Ritter. Categorical and Kripke semantics for constructive S4 modal logic. In Laurent Fribourg, editor, *Computer Science Logic, 15th International Workshop, CSL 2001, 10th Annual Conference of the EACSL, Paris, France, September 10–13, 2001, Proceedings*, volume 2142 of *Lecture Notes in Computer Science*, pages 292–307. Springer, 2001.

[2] Ryuta Arisaka, Anupam Das, and Lutz Straßburger. On nested sequents for constructive modal logics. *Logical Methods in Computer Science*, 11(3:7), 2015.

[3] Sergei Artemov and Melvin Fitting. Justification logic. In Edward N. Zalta, editor, *The Stanford Encyclopedia of Philosophy*. Metaphysics Research Lab, Stanford University, winter 2016 edition, 2016.

[4] Sergei Artemov and Melvin Fitting. *Justification Logic: Reasoning with Reasons*, volume 216 of *Cambridge Tracts in Mathematics*. Cambridge University Press, 2019.

[5] Sergei Artemov and Rosalie Iemhoff. The basic intuitionistic logic of proofs. *Journal of Symbolic Logic*, 72(2):439–451, 2007.

[6] Sergei N. Artemov. Operational modal logic. Technical Report MSI 95–29, Cornell University, 1995.

[7] Sergei N. Artemov. Explicit provability and constructive semantics. *Bulletin of Symbolic Logic*, 7(1):1–36, 2001.

[8] Sergei N. Artemov. Unified semantics for modality and λ-terms via proof polynomials. In Kees Vermeulen and Ann Copestake, editors, *Algebras, Diagrams and Decisions in Language, Logic and Computation*, volume 144 of *CSLI Lecture Notes*, pages 89–118. CSLI Publications, 2002.

[9] Sergei [N.] Artemov. The logic of justification. *Review of Symbolic Logic*, 1(4):477–513, 2008.

[10] Sergei N. Artemov. The ontology of justifications in the logical setting. *Studia Logica*, 100(1–2):17–30, 2012.

[11] Sergei N. Artemov, E. L. Kazakov, and D. Shapiro. Logic of knowledge with justifications. Technical Report CFIS 99–12, Cornell University, 1999.

[12] G. M. Bierman and V. C. V. de Paiva. On an intuitionistic modal logic. *Studia Logica*, 65(3):383–416, 2000.

[13] Vladimir N. Brezhnev. On explicit counterparts of modal logics. Technical Report CFIS 2000–05, Cornell University, 2000.

[14] Samuel Bucheli, Roman Kuznets, and Thomas Studer. Realizing public announcements by justifications. *Journal of Computer and System Sciences*, 80(6):1046–1066, September 2014.

[15] Kaustuv Chaudhuri, Sonia Marin, and Lutz Straßburger. Modular focused proof systems for intuitionistic modal logics. In Delia Kesner and Brigitte Pientka, editors, *1st International Conference on Formal Structures for Computation and Deduction, FSCD 2016, June 22–26, 2016, Porto, Portugal*, volume 52 of *Leibniz International Proceedings in Informatics (LIPIcs)*. Schloss Dagstuhl–Leibniz-Zentrum für Informatik,

2016.

[16] Evgenij Dashkov. Arithmetical completeness of the intuitionistic logic of proofs. *Journal of Logic and Computation*, 21(4):665–682, 2011.

[17] Melvin Fitting. The logic of proofs, semantically. *Annals of Pure and Applied Logic*, 132(1):1–25, 2005.

[18] Melvin Fitting. Modal logics, justification logics, and realization. *Annals of Pure and Applied Logic*, 167(8):615–648, August 2016.

[19] Melvin Fitting. Realization using the model existence theorem. *Journal of Logic and Computation*, 26(1):213–234, February 2016.

[20] Kurt Gödel. Eine Interpretation des intuitionistischen Aussagenkalküls. *Ergebnisse eines mathematischen Kolloquiums*, 4:39–40, 1933.

[21] Kurt Gödel. Vortrag bei Zilsel/Lecture at Zilsel's (*1938a). In Solomon Feferman, John W. Dawson, Jr., Warren Goldfarb, Charles Parsons, and Robert M. Solovay, editors, *Unpublished essays and lectures*, volume III of *Kurt Gödel Collected Works*, pages 86–113. Oxford University Press, 1995.

[22] Remo Goetschi and Roman Kuznets. Realization for justification logics via nested sequents: Modularity through embedding. *Annals of Pure and Applied Logic*, 163(9):1271–1298, 2012.

[23] Roman Kuznets and Thomas Studer. Justifications, ontology, and conservativity. In Thomas Bolander, Torben Braüner, Silvio Ghilardi, and Lawrence Moss, editors, *Advances in Modal Logic, Volume 9*, pages 437–458. College Publications, 2012.

[24] Roman Kuznets and Thomas Studer. *Logics of Proofs and Justifications*, volume 80 of *Studies in Logic*. College Publications, 2019.

[25] Björn Lellmann and Dirk Pattinson. Constructing cut free sequent systems with context restrictions based on classical or intuitionistic logic. In Kamal Lodaya, editor, *Logic and Its Applications, 5th Indian Conference, ICLA 2013, Chennai, India, January 10–12, 2013, Proceedings*, volume 7750 of *Lecture Notes in Computer Science*, pages 148–160. Springer, 2013.

[26] Michel Marti and Thomas Studer. Intuitionistic modal logic made explicit. *IfCoLog Journal of Logics and their Applications*, 3(5):877–901, 2016.

[27] Michel Marti and Thomas Studer. The internalized disjunction property for intuitionistic justification logic. In Guram Bezhanishvili, Giovanna D'Agostino, George Metcalfe, and Thomas Studer, editors, *Advances in Modal Logic, Volume 12*, pages 511–529. College Publications, 2018.

[28] Michael Mendler and Valeria de Paiva. Constructive CK for contexts. In Luciano Serafini and Paolo Bouquet, editors, *Proceedings of the CRR '05 Workshop on Context Representation and Reasoning, Paris, France, July 5, 2005*, volume 136 of *CEUR Workshop Proceedings*. Luciano Serafini and Paolo Bouquet, 2005.

[29] Michael Mendler and Stephan Scheele. Cut-free Gentzen calculus for multimodal CK. *Information and Computation*, 209(12):1465–1490, 2011.

[30] Alex Simpson. *The Proof Theory and Semantics of Intuitionistic Modal Logic*. PhD

thesis, University of Edinburgh, 1994.

[31] Gabriela Steren and Eduardo Bonelli. Intuitionistic hypothetical logic of proofs. In Valeria de Paiva, Mario Benevides, Vivek Nigam, and Elaine Pimentel, editors, *Proceedings of the 6th Workshop on Intuitionistic Modal Logic and Applications (IMLA 2013)*, volume 300 of *Electronic Notes in Theoretical Computer Science*, pages 89–103. Elsevier, 2014.

[32] Lutz Straßburger. Cut elimination in nested sequents for intuitionistic modal logics. In Frank Pfenning, editor, *Foundations of Software Science and Computation Structures, 16th International Conference, FOSSACS 2013, Held as Part of the European Joint Conferences on Theory and Practice of Software, ETAPS 2013, Rome, Italy, March 16–24, 2013, Proceedings*, volume 7794 of *Lecture Notes in Computer Science*, pages 209–224. Springer, 2013.

[33] A. S. Troelstra and H. Schwichtenberg. *Basic Proof Theory*, volume 43 of *Cambridge Tracts in Theoretical Computer Science*. Cambridge University Press, 2nd edition, 2000.

[34] Heinrich Wansing. Sequent systems for modal logics. volume 8 of *Handbook of Philosophical Logic, 2nd Edition*, pages 61–145. Kluwer Academic Publishers, 2002.

[35] Duminda Wijesekera. Constructive modal logics I. *Annals of Pure and Applied Logic*, 50(3):271–301, 1990.

Received 11 October 2017

Glivenko's Theorem, Finite Height, and Local Tabularity

Ilya B. Shapirovsky
New Mexico State University, USA
and
Institute for Information Transmission Problems of Russian Academy of Sciences
ilshapir@nmsu.edu

Abstract

Glivenko's theorem states that a formula is derivable in classical propositional logic CL iff under the double negation it is derivable in intuitionistic propositional logic IL: $CL \vdash \varphi$ iff $IL \vdash \neg\neg\varphi$. Its analog for the modal logics S5 and S4 states that $S5 \vdash \varphi$ iff $S4 \vdash \neg\Box\neg\Box\varphi$. In Kripke semantics, IL is the logic of partial orders, and CL is the logic of partial orders of height 1. Likewise, S4 is the logic of preorders, and S5 is the logic of equivalence relations, which are preorders of height 1. In this paper we generalize Glivenko's translation for logics of arbitrary finite height.

Keywords: Glivenko's translation, modal logic, intermediate logic, finite height, pretransitive logic, local tabularity, local finiteness, top-heavy frame

1 Introduction

For a modal or intermediate logic L, let L[h] be its extension with the formula restricting the height of a Kripke frame by finite h. In the intermediate case, such formulas are defined as $B_0^I = \bot$, $B_h^I = p_h \vee (p_h \to B_{h-1}^I)$, and in the modal transitive case as $B_0 = \bot$, $B_h = p_h \to \Box(\Diamond p_h \vee B_{h-1})$. In particular, classical logic CL is the extension of intuitionistic logic IL with the formula $p \vee \neg p$, that is $CL = IL[1]$.

I would like to thank Valentin Shehtman for many useful discussions and comments. I would also like to thank the three anonymous reviewers for their suggestions and questions on the early version of the manuscript.

The work on this paper was supported by the Russian Science Foundation under grant 16-11-10252 and carried out at Steklov Mathematical Institute of Russian Academy of Sciences.

Similarly, S5 = S4[1]. Glivenko's translation [13] and its analog for the modal logics S5 and S4 [20] can be formulated as follows:

$$\text{IL}[1] \vdash \varphi \quad \text{iff} \quad \text{IL} \vdash \neg\neg\varphi, \tag{1}$$

$$\text{S4}[1] \vdash \varphi \quad \text{iff} \quad \text{S4} \vdash \Diamond\Box\varphi. \tag{2}$$

For finite variable fragments of IL and S4, the above equivalences can be generalized for arbitrary finite height. A *k-formula* is a formula in variables $p_0, \ldots p_{k-1}$. Let (W, R) be the k-generated canonical frame of S4 (that is, W is the set of maximal S4-consistent sets of k-formulas). It follows from [24] (see also [25], [9], [1], [2]) that there exist formulas $\mathbf{B}_{h,k}$ (and their intuitionistic analogs $\mathbf{B}^{\text{I}}_{h,k}$) such that for every $x \in W$, $\mathbf{B}_{h,k} \in x$ iff the depth of x in W is less than or equal to h. We observe that for all finite k, for all k-formulas φ,

$$\text{IL}[h+1] \vdash \varphi \quad \text{iff} \quad \text{IL} \vdash \bigwedge_{i \leq h} ((\varphi \to \mathbf{B}^{\text{I}}_{i,k}) \to \mathbf{B}^{\text{I}}_{i,k}), \tag{3}$$

$$\text{S4}[h+1] \vdash \varphi \quad \text{iff} \quad \text{S4} \vdash \bigwedge_{i \leq h} (\Box(\Box\varphi \to \mathbf{B}_{i,k}) \to \mathbf{B}_{i,k}). \tag{4}$$

In particular, for $h = 0$ we have equivalences (1) and (2), since the formulas $\mathbf{B}_{0,k}$ and $\mathbf{B}^{\text{I}}_{0,k}$ are $\bot$ for all k.

Sometimes, analogs of the formulas $\mathbf{B}_{h,k}$ exist for unimodal logics smaller than S4 and for polymodal logics. A modal logic L is *pretransitive* (or *weakly transitive*, in another terminology), if the transitive reflexive closure modality $\Diamond^*$ is expressible in L [15]. Namely, for the language with n modalities $\Diamond_i$ ($i < n$), put $\Diamond^0\varphi = \varphi$, $\Diamond^{m+1}\varphi = \Diamond^m \bigvee_{i<n} \Diamond_i\varphi$, $\Diamond^{\leq m}\varphi = \bigvee_{l \leq m}\Diamond^l\varphi$. A logic L is *pretransitive* if it contains $\Diamond^{m+1}p \to \Diamond^{\leq m}p$ for some finite m. In this case $\Diamond^{\leq m}$ plays the role of $\Diamond^*$. The *height of a polymodal frame* $(W, (R_i)_{i<n})$ is the height of the preorder $(W, (\bigcup_{i<n} R_i)^*)$. In the pretransitive case, formulas of finite height can be defined analogously to the transitive case.

L is said to be *k-tabular* if, up to the equivalence in L, there exist only finitely many k-formulas. L is *locally tabular* (or *locally finite*) if it is k-tabular for every finite k.

We show (Theorems 5 and 6) that if L is a pretransitive logic, $h, k < \omega$, and L[h] is k-tabular, then:

1. For every $i \leq h$, there exists a formula $\mathbf{B}_{i,k}$ such that $\mathbf{B}_{i,k} \in x$ iff the depth of x in the k-generated canonical frame of L is less than or equal to i.

2. For all k-formulas φ,

$$\text{L}[h+1] \vdash \varphi \text{ iff } \text{L} \vdash \bigwedge_{i \leq h} (\Box^*(\Box^*\varphi \to \mathbf{B}_{i,k}) \to \mathbf{B}_{i,k}). \tag{5}$$

The equivalence (5) generalizes (4). Recall that a unimodal transitive logic is locally tabular iff it is of finite height iff it is 1-tabular ([22], [18]). In the non-transitive case the situation is much more complicated. It follows from [23] that every locally tabular (even 1-tabular) logic is a pretransitive logic of finite height; however, it follows from [17] that there exists a pretransitive L such that none of the logics L[h] are 1-tabular ($h > 0$). In Section 5 we discuss how k-tabularity of L[h] depends on h and k. In particular, we construct the first example of a modal logic which is 1-tabular but not locally tabular.

2   Preliminaries

Fix a finite $n > 0$; *n-modal formulas* are built from a countable set $\{p_0, p_1, \ldots\}$ of proposition letters, the classical connectives $\to$, $\bot$, and the modal connectives $\Diamond_i$, $i < n$; the other Boolean connectives are defined as standard abbreviations; $\Box_i$ abbreviates $\neg\Diamond_i\neg$. We omit the subscripts on the modalities when $n = 1$. By a *logic* we mean a *propositional n-modal normal logic*, that is a set of n-modal formulas containing all classical tautologies, the formulas $\Diamond_i(p \vee q) \to \Diamond_i p \vee \Diamond_i q$ and $\neg\Diamond_i\bot$ for all $i < n$, and closed under the rules of Modus Ponens, Substitution, and Monotonicity (if $\varphi \to \psi$ is in the logic, then so is $\Diamond_i\varphi \to \Diamond_i\psi$).

For a logic L and a set of formulas Ψ, the smallest logic containing $L \cup \Psi$ is denoted by $L + \Psi$. For a formula φ, the notation $L + \varphi$ abbreviates $L + \{\varphi\}$. In particular, $K4 = K + \Diamond\Diamond p \to \Diamond p$, $S4 = K4 + p \to \Diamond p$, $S5 = S4 + p \to \Diamond\Box p$, where K denotes the smallest unimodal logic. $L \vdash \varphi$ is a synonym for $\varphi \in L$.

The truth and the validity of modal formulas in Kripke frames and models are defined as usual, see, e.g., [6]. By a *frame* we always mean a Kripke frame $(W, (R_i)_{i<n})$, $W \neq \emptyset$, $R_i \subseteq W \times W$. We put $R_F = \cup_{i<n} R_i$. The transitive reflexive closure of a relation R is denoted by R^*; the notation $R(x)$ is used for the set $\{y \mid xRy\}$. The *restriction* of F onto its subset V, $F{\upharpoonright}V$ in symbols, is $(V, (R_i \cap (V \times V))_{i<n})$. In particular, we put $F\langle x\rangle = F{\upharpoonright}R_F^*(x)$.

For $k \leq \omega$, a *k-formula* is a formula in proposition letters p_i, $i < k$.

Let L be a consistent logic. For $k \leq \omega$, the *k-canonical model of* L is built from maximal L-consistent sets of k-formulas; the relations and the valuation are defined in the standard way, see e.g. [8]. Recall the following fact.

Proposition 1 (Canonical model theorem). Let M be the k-canonical model of a logic L, $k \leq \omega$. Then for all k-formulas φ we have:

1. $M, x \vDash \varphi$ iff $\varphi \in x$, for all x in M;

2. $M \vDash \varphi$ iff $L \vdash \varphi$.

A logic L is said to be *k-tabular* if, up to the equivalence in L, there exist only finitely many k-formulas. L is *locally tabular* (or *locally finite*) if it is k-tabular for every finite k. The following proposition is straightforward from the definitions.

Proposition 2. Let L be a logic, $k < \omega$. The following are equivalent:

- L is k-tabular.

- The k-generated Lindenbaum-Tarski algebra of L is finite.

- The k-canonical model of L is finite.

A unimodal logic is *transitive* if it contains the formula $\Diamond\Diamond p \to \Diamond p$; recall that this formula expresses transitivity of a binary relation. Below we consider a weaker property, *pretransitivity* of (polymodal) logics and frames.

For a binary relation R on a set W, put $R^{\leq m} = \cup_{i \leq m} R^i$, where $R^0 = Id(W)$, $R^{i+1} = R \circ R^i$. R is called *m-transitive* if $R^{\leq m} = R^*$. R is *pretransitive* if it is m-transitive for some m. A frame F is *m-transitive* if R_F is m-transitive.

Let $\Diamond^0 \varphi = \varphi$, $\Diamond^{i+1}\varphi = \Diamond^i(\Diamond_0\varphi \vee \ldots \vee \Diamond_{n-1}\varphi)$, $\Diamond^{\leq m}\varphi = \vee_{i \leq m}\Diamond^i\varphi$, $\Box^{\leq m}\varphi = \neg\Diamond^{\leq m}\neg\varphi$.

Proposition 3. Let F be a frame. The following are equivalent:

- F is m-transitive;

- $R_F^{m+1} \subseteq R_F^{\leq m}$;

- $F \vDash \Diamond^{m+1}p \to \Diamond^{\leq m}p$.

The proof is straightforward, details can be found, e.g., in [15].

Definition 1. A logic L is said to be *m-transitive* if $L \vdash \Diamond^{m+1}p \to \Diamond^{\leq m}p$. L is *pretransitive* if it is m-transitive for some $m \geq 0$.[1]

If L is pretransitive, then there exists the least m such that L is m-transitive; in this case we write $\Diamond^*\varphi$ for $\Diamond^{\leq m}\varphi$, and $\Box^*\varphi$ for $\Box^{\leq m}\varphi$.

For a unimodal formula φ, $\varphi^{[*]}$ denotes the formula obtained from φ by replacing $\Diamond$ with $\Diamond^*$ and $\Box$ with $\Box^*$.

Proposition 4. For a pretransitive logic L, the set $\{\varphi \mid L \vdash \varphi^{[*]}\}$ is a logic containing S4.

---

[1]Pretransitive logics sometimes are called *weakly transitive*. However, in the other terminology, the term 'weakly transitive' is used for logics containing the formula $\Diamond\Diamond p \to \Diamond p \vee p$.

Proof. Follows from [12, Lemma 1.3.45]. $\qquad\qquad\qquad\qquad\qquad\qquad\qquad$ $\square$

A poset is of *height* $h < \omega$ if it contains a chain of h elements and no chains of cardinality $> h$.

A *cluster* in a frame F is an equivalence class with respect to the relation $\sim_F = R_F^* \cap R_F^{*-1}$. For clusters C, D, put $C \leq_F D$ iff xR_F^*y for some $x \in C, y \in D$. The poset $(W/\sim_F, \leq_F)$ is called the *skeleton of* F. The *height of a frame* F, in symbols $\mathrm{ht}(F)$, is the height of its skeleton.

Put

$$B_0 = \bot, \quad B_{i+1} = p_{i+1} \to \Box^*(\Diamond^* p_{i+1} \vee B_i).$$

In the unimodal transitive case, the formula B_h expresses the fact that the height of a frame $\leq h$ [22]. In the case when $F = (W, (R_i)_{i<n})$ is m-transitive, the operator $\Diamond^* = \Diamond^{\leq m}$ relates to R_F^*. Since the height of F is the height of the preorder (W, R_F^*), we have $F \vDash B_h$ iff $\mathrm{ht}(F) \leq h$.

Definition 2. A pretransitive logic is of *finite height* if it contains B_h for some $h < \omega$. For a pretransitive L, we put

$$L[h] = L + B_h.$$

Example 1. Unimodal examples of 1-transitive logics are S4, $\mathrm{wK4} = K + \Diamond\Diamond p \to \Diamond p \vee p$. The logic S5 and the *difference logic* $\mathrm{DL} = \mathrm{wK4} + p \to \Box\Diamond p$ are examples of logics of height 1.

A well-known logic $\mathrm{K5} = K + \Diamond p \to \Box\Diamond p$ is a 2-transitive logic of height 2. To show this, recall that K5 is Kripke complete and its frames are those that validate the sentence $\forall x \forall y \forall z (xRy \wedge xRz \to yRz)$. Every K5-frame is 2-transitive. Indeed, suppose that $aRbRcRd$ for some elements of a K5-frame. Then bRb; we also have bRc, thus cRb; from cRb and cRd we infer that bRd. Thus aR^2d. It is not difficult to see that if a K5-frame F has an irreflexive serial point, then the height of F is 2; otherwise F is a disjoint sum of S5-frames and irreflexive singletons, so its height is 1.

Theorem 1 ([22, 18]). *A unimodal transitive logic is locally tabular iff it is of finite height.*

In [23], it was shown that every locally tabular unimodal logic is a pretransitive logic of finite height; in fact, the proof yields the following stronger formulation.

Theorem 2. *If a logic is 1-tabular, then it is a pretransitive logic of finite height.*

Proof. Let L be 1-tabular. Then its 1-canonical frame is finite. Every finite frame is m-transitive for some m. Thus L is m-transitive.

By Proposition 4, the set $*L = \{\varphi \mid L \vdash \varphi^{[*]}\}$ is a logic containing S4. Since L is 1-tabular, $*L$ is 1-tabular. In [18], it was shown that for transitive logics 1-tabularity implies local tabularity. Thus $*L$ is of finite height. It follows that L is of finite height too. ☐

Thus, all locally tabular logics are pretransitive of finite height. However, unlike the transitive case, the converse is not true in general even for unimodal logics. Let TR_m be the smallest m-transitive unimodal logic. For $m \geq 2$, $h \geq 1$, none of the logics $TR_m[h]$ are locally tabular [7]; moreover, they are not 1-tabular [17].

3 Translation for logics of height 1

For a pretransitive logics L, $L[1] = L + B_1$, that is $L[1]$ is the smallest logic containing $L \cup \{p \to \Box^*\Diamond^*p\}$. It is known that $S4[1] = S5 \vdash \varphi$ iff $S4 \vdash \Diamond\Box\varphi$, and $S5 \vdash \Box\psi \to \Box\varphi$ iff $S4 \vdash \Diamond\Box\psi \to \Diamond\Box\varphi$ [20], [21]. In [16], it was shown that in the pretransitive unimodal case we have $L[1] \vdash \varphi$ iff $L \vdash \Diamond^*\Box^*\varphi$. In this section we generalize these facts to the polymodal case using the maximality property of pretransitive canonical frames (see Proposition 6 below).

Proposition 5. Let F be the k-canonical frame of a pretransitive logic L, $k \leq \omega$. For all x, y in F, we have

$$xR_F^*y \text{ iff } \forall\varphi \, (\varphi \in y \Rightarrow \Diamond^*\varphi \in x).$$

The proof is straightforward; for details see, e.g., Proposition 5.9 and Theorem 5.16 in [8].

Consider a frame F and its subset V. We say that $x \in V$ is a *maximal element* of V, if for all $y \in V$, xR_F^*y implies yR_F^*x.

It is known that in canonical transitive frames every non-empty definable subset has a maximal element [9]; the next proposition shows that this property holds in the pretransitive case as well.

Proposition 6 (Maximality lemma). Suppose that F is the k-canonical frame of a pretransitive L, $k \leq \omega$. Let $\varphi \in x$ for some x in F and some formula φ. Then $R_F^*(x) \cap \{y \mid \varphi \in y\}$ has a maximal element.

Proof. For a formula α, put $\|\alpha\| = \{y \mid \alpha \in y\}$. Since $\varphi \in x$, $R_F^*(x) \cap \|\varphi\|$ is non-empty.

2338

Let Σ be an R_{F}^*-chain in $R_{\mathrm{F}}^*(x) \cap \|\varphi\|$. The family $\{R_{\mathrm{F}}^*(y) \cap \|\varphi\| \mid y \in \Sigma\}$ has the finite intersection property (indeed, if Σ_0 is a non-empty finite subset of Σ, then for some $y_0 \in \Sigma_0$ we have $y R_{\mathrm{F}}^* y_0$ for all $y \in \Sigma_0$; so $y_0 \in R_{\mathrm{F}}^*(y) \cap \|\varphi\|$ for all $y \in \Sigma_0$). By Proposition 5, $R_{\mathrm{F}}^*(y) = \bigcap\{\|\alpha\| \mid \Box^* \alpha \in y\}$. It follows that all sets $R_{\mathrm{F}}^*(y) \cap \|\varphi\|$ are closed in the Stone topology on F (see, e.g., [14, Theorem 1.9.4]). By the compactness, $\bigcap\{R_{\mathrm{F}}^*(y) \cap \|\varphi\| \mid y \in \Sigma\}$ is non-empty. Thus Σ has an upper bound in $\|\varphi\|$. By Zorn's lemma, $R_{\mathrm{F}}^*(x) \cap \|\varphi\|$ contains a maximal element. $\quad\square$

Proposition 7. A pretransitive logics L is consistent iff L[1] is consistent.

Proof. Easily follows from Proposition 4 and the fact that if a logic containing S4 is consistent, then its extension with the formula $p \to \Box\Diamond p$ is consistent. $\quad\square$

Since $L[1] \supseteq L[2] \supseteq L[3] \supseteq \ldots$, it follows that if L is consistent, then L[h] is consistent for any $h > 0$.

For a frame F and a point x is F, the *depth of x in* F is the height of the frame $F\langle x \rangle$. Let F[h] denote the restriction of F onto the set of its points of depth less than or equal to h, i.e., $F[h] = F{\restriction}\{x \mid \mathrm{ht}(F\langle x \rangle) \leq h\}$.

Proposition 8. Let F be the k-canonical frame of a pretransitive logic L, $k \leq \omega$.

1. For all x in F, $0 \leq h < \omega$,
$$\text{the depth of } x \text{ in } F \text{ is } \leq h \quad \text{iff}$$
$$B_h(\psi_1, \ldots, \psi_h) \in x \text{ for all } k\text{-formulas } \psi_1, \ldots, \psi_h.$$

2. For $0 < h < \omega$, the frame F[h] is the canonical frame of L[h].

Proof. 1. If $\mathrm{ht}(F\langle x \rangle) \leq h$, then B_h is valid at x in F; by the Canonical model theorem, $B_h(\psi_1, \ldots, \psi_h) \in x$ for all k-formulas $\psi_1, \ldots, \psi_h$.

By induction on h, let us show that if $\mathrm{ht}(F\langle x \rangle) > h$, then $B_h(\psi_1, \ldots, \psi_h) \notin x$ for some $\psi_1, \ldots, \psi_h$. The basis is trivial, since there are no points containing $B_0 = \bot$ in F. Suppose $\mathrm{ht}(F\langle x \rangle) > h + 1$. Then there exists y such that $\mathrm{ht}(F\langle y \rangle) > h$, $(x, y) \in R_{\mathrm{F}}^*$, and $(y, x) \notin R_{\mathrm{F}}^*$. By induction hypothesis, $B_h(\psi_1, \ldots, \psi_h) \notin y$ for some $\psi_1, \ldots, \psi_h$. By Proposition 5, for some ψ_{h+1} we have $\psi_{h+1} \in x$ and $\Diamond^* \psi_{h+1} \notin y$. It follows that $B_{h+1}(\psi_1, \ldots, \psi_{h+1}) \notin x$.

2. Since $L \subseteq L[h]$, the k-canonical frame of L[h] is a generated subframe of F. Now the statement follows from the first statement of the proposition. $\quad\square$

A logic L is k-*canonical* if it is valid in its k-canonical frame.

Proposition 9. If a pretransitive L is k-canonical $(k \leq \omega)$, then L[h] is k-canonical for all $0 < h < \omega$.

Proof. Follows from Proposition 8. ☐

Theorem 3. Let L be a pretransitive logic. Then for all formulas φ, ψ we have

$$L[1] \vdash \Box^*\psi \to \Box^*\varphi \text{ iff } L \vdash \Diamond^*\Box^*\psi \to \Diamond^*\Box^*\varphi.$$

Proof. By Proposition 7, we may assume that both L and L[1] are consistent. Let F be the ω-canonical frame of L, and G the ω-canonical frame of L[1].

Suppose $L \vdash \Diamond^*\Box^*\psi \to \Diamond^*\Box^*\varphi$. Consider an element x of G. By Proposition 4, $\{\varphi \mid L[1] \vdash \varphi^{[*]}\}$ is a logic containing S5. Thus x contains formulas $\Box^*\psi \to \Diamond^*\Box^*\psi$ and $\Diamond^*\Box^*\varphi \to \Box^*\varphi$. Since $L \subseteq L[1]$, x also contains $\Diamond^*\Box^*\psi \to \Diamond^*\Box^*\varphi$. It follows that if x contains $\Box^*\psi$, then x contains $\Box^*\varphi$. By the Canonical model theorem, $L[1] \vdash \Box^*\psi \to \Box^*\varphi$.

Now suppose $L[1] \vdash \Box^*\psi \to \Box^*\varphi$. Assume that $\Diamond^*\Box^*\psi \in x$ for some element x of F. Then for some y we have $\Box^*\psi \in y$ and xR_F^*y. The set $R_F^*(y)$ has a maximal element z by Proposition 6. It follows that $ht(F\langle z\rangle) = 1$. By Proposition 8, $G = F[1]$. Thus z is in G and hence $\Box^*\psi \to \Box^*\varphi$ is in z. Since yR_F^*z, we have $\Box^*\psi \in z$, which implies that $\Box^*\varphi \in z$. Hence $\Diamond^*\Box^*\varphi$ is in x. It follows that $L \vdash \Diamond^*\Box^*\psi \to \Diamond^*\Box^*\varphi$. ☐

Theorem 4. Let L be a pretransitive logic.

1. For all φ, we have $L[1] \vdash \varphi$ iff $L \vdash \Diamond^*\Box^*\varphi$.

2. If L is decidable, then so is L[1].

3. If L has the finite model property, then so does L[1].

Proof. By Theorem 3, we have

$$L[1] \vdash \Box^*\top \to \Box^*\varphi \text{ iff } L \vdash \Diamond^*\Box^*\top \to \Diamond^*\Box^*\varphi.$$

By Proposition 4, we have $\top \leftrightarrow \Box^*\top$ and $\top \leftrightarrow \Diamond^*\Box^*\top$ in every pretransitive logic; also, we have $\Box^*\varphi \in L[1]$ iff $\varphi \in L[1]$. Now the first statement follows.

The second statement is an immediate consequence of the first one.

Suppose L has the finite model property. Consider a formula $\varphi \notin L[1]$. Then $\Diamond^*\Box^*\varphi \notin L$. Then $\Diamond^*\Box^*\varphi$ is refuted in some finite L-frame F. If follows that φ is refuted in F at some point in a maximal cluster C. The restriction $F{\upharpoonright}C$ is a generated subframe of F. Thus $F{\upharpoonright}C$ refutes φ and validates L. The height of this restriction is 1, so $F{\upharpoonright}C \vDash L[1]$. Thus L[1] has the finite model property. ☐

Example 2. Important examples of pretransitive frames are birelational frames $(W, \leq, R)$ with transitive R. Recall that $(W, \leq, R)$ is a *birelational frame*, if $\leq$ is a partial order on W, $R \subseteq W^2$, and

$$(R \circ \leq) \subseteq (\leq \circ R), \quad (R^{-1} \circ \leq) \subseteq (\leq \circ R^{-1}).$$

Consider the class of all birelational frames $(W, \leq, R)$ with transitive reflexive R. Its modal logic L is the smallest bimodal logic containing the axioms of S4 for modalities $\Box_0, \Box_1$, and the formulas $\Diamond_1 \Diamond_0 p \rightarrow \Diamond_0 \Diamond_1 p$ and $\Diamond_0 \Box_1 p \rightarrow \Box_1 \Diamond_0 p$ [11] (recall that in the semantics of modal intuitionistic logic, the logic of this class is known to be IS4 [10], one of the *"most prominent logics for which decidability is left open"* [26]). In this case, $\Diamond_0 \Diamond_1$ plays the role of the master modality, and the formula B_1 says that $\leq \circ R$ is an equivalence. The decidability and the finite model property of the logic L, as well as of the logic IS4, is an open question. By the above theorem, we have $L \vdash \Diamond_0 \Diamond_1 \Box_0 \Box_1 \varphi$ iff $L[1] \vdash \varphi$.

Question. Is the logic L[1] decidable? Does it have the fmp?

4 Translation for logics of arbitrary finite height

In the proof of Theorem 3 we used the following property of a canonical frame F of L: every point in F is below (w.r.t. to the preorder R_F^*) a maximal point; maximal points form F[1], the canonical frame of L[1]. To describe translations from L[h] to L for $h > 1$, we shall use the following analog of this property.

Definition 3. Let $0 < h < \omega$. A frame F is said to be *h-heavy* if for its every element x which is not in F[h] there exists y such that $x R_F^* y$ and $\mathrm{ht}(F\langle y \rangle) = h$.

F is said to be *top-heavy* if it is h-heavy for all positive finite h.

Proposition 10. The k-canonical frame of a consistent pretransitive logic is 1-heavy for every $k \leq \omega$.

Proof. In the Maximality lemma (Proposition 6), put $\varphi = \top$. □

It is known that k-canonical frames of unimodal transitive logics are top-heavy for all finite k ([24], [9], [1]).[2] This can be generalized for the pretransitive case as follows.

Theorem 5. Let L be a consistent pretransitive logic, $h, k < \omega$. If L[h] is k-tabular, then:

---

[2]The term 'top-heavy' was introduced in [9].

1. For every $i \leq h$, there exists a formula $\mathbf{B}_{i,k}$ such that $\mathbf{B}_{i,k} \in x$ iff the depth of x in the k-canonical frame of L is less than or equal to i.

2. The k-canonical frame of L is $(h+1)$-heavy.

Proof. The case $h = 0$ follows from Proposition 10. Suppose $h > 0$.

Let $\mathrm{F} = (W, (R_i)_{i<n})$ be the k-canonical frame of L. By Proposition 8, the frame $\mathrm{F}[h] = (\overline{W}, (\overline{R}_i)_{i<n})$ is the k-canonical frame of $\mathrm{L}[h]$. Since $\mathrm{L}[h]$ is k-tabular, it follows that $\overline{W}$ is finite and for every a in $\overline{W}$ there exists a k-formula $\alpha(a)$ such that

$$\forall b \in \overline{W} \, (\alpha(a) \in b \iff b = a). \tag{6}$$

Without loss of generality we may assume that $\alpha(a)$ is of the form

$$p_0^{\pm} \wedge \ldots \wedge p_{k-1}^{\pm} \wedge \varphi, \tag{7}$$

where $p_i^{\pm} \in \{p_i, \neg p_i\}$.

For $a \in \overline{W}$ let $\beta(a)$ be the following Jankov-Fine formula:

$$\beta(a) = \alpha(a) \wedge \gamma, \tag{8}$$

where γ is the conjunction of the formulas

$$\Box^* \bigwedge \left\{ \alpha(b_1) \to \Diamond_i \alpha(b_2) \mid (b_1, b_2) \in \overline{R}_i, \, i < n \right\} \tag{9}$$

$$\Box^* \bigwedge \left\{ \alpha(b_1) \to \neg \Diamond_i \alpha(b_2) \mid (b_1, b_2) \in \overline{W}^2 \setminus \overline{R}_i, \, i < n \right\} \tag{10}$$

$$\Box^* \bigvee \left\{ \alpha(b) \mid b \in \overline{W} \right\} \tag{11}$$

For all $x, y \in W$, $i < n$ we have

$$\text{if } \gamma \in x \text{ and } x R_i y, \text{ then } \gamma \in y. \tag{12}$$

We claim that

$$\forall a \in \overline{W} \, \forall x \in W \, (\beta(a) \in x \iff x = a). \tag{13}$$

To prove this, by induction on the length of formulas we show that for all k-formulas φ, all $a \in \overline{W}$, and all $x \in W$,

$$\text{if } \beta(a) \in x, \text{ then } \varphi \in a \iff \varphi \in x. \tag{14}$$

The basis of induction follows from (7). The Boolean cases are trivial.

Assume that $\varphi = \Diamond_i \psi$.

First, suppose $\Diamond_i\psi \in a$. We have $\psi \in b$ for some b with $a\overline{R}_i b$. Since $\beta(a) \in x$, by (9) we have $\Diamond_i\alpha(b) \in x$. Then we have $\alpha(b) \in y$ for some y with xR_iy; by (12), $\beta(b) \in y$. Hence $\psi \in y$ by induction hypothesis. Thus $\Diamond_i\psi \in x$.

Now let us show that $\Diamond_i\psi \in a$ whenever $\Diamond_i\psi \in x$. In this case we have $\psi \in y$ for some y with xR_iy. By (11) we infer that $\alpha(b) \in y$ for some $b \in \overline{W}$. Thus $\Diamond_i\alpha(b) \in x$. Since $\alpha(a) \in x$, it follows from (10) that $a\overline{R}_ib$. By (12) we have $\gamma \in y$, thus $\beta(b) \in y$; by induction hypothesis $\psi \in b$. Hence $\Diamond_i\psi \in a$, as required.

Thus (14) is proved and (13) follows.

Now using the formulas (8), for $i \leq h$ we can define the formulas $\mathbf{B}_{i,k}$ such that for all x in F,

$$\text{the depth of } x \text{ in F is } \leq i \text{ iff } \mathbf{B}_{i,k} \in x. \tag{15}$$

For this, we put

$$\mathbf{B}_{i,k} = \bigvee\{\beta(a) \mid \operatorname{ht}(\mathrm{F}\langle a\rangle) \leq i\}. \tag{16}$$

This proves the first statement of the theorem.

In particular, it follows that $W \setminus \overline{W}$ is definable in the k-canonical model of L:

$$x \in W \setminus \overline{W} \text{ iff } \neg\mathbf{B}_{h,k} \in x.$$

Now by Proposition 6 we have that if x is not in $\overline{W}$, then there exists a maximal y in $R_{\mathrm{F}}^*(x) \setminus \overline{W}$. Hence if $(y, z) \in R_{\mathrm{F}}^*$ and $(z, y) \notin R_{\mathrm{F}}^*$ for some z, then z belongs to $\overline{W}$, which means $\operatorname{ht}(\mathrm{F}\langle z\rangle) \leq h$. Thus $\operatorname{ht}(\mathrm{F}\langle y\rangle) \leq h + 1$. On the other hand, $y \notin \overline{W}$. It follows that $\operatorname{ht}(\mathrm{F}\langle y\rangle) = h + 1$, as required. □

The logic L[0] is inconsistent, so it is k-tabular. Hence Proposition 10 can be considered as a particular case of the above theorem.

Note that formulas (8) define atoms in the Lindenbaum-Tarski (i.e., free) k-generated algebra of L.

Theorem 6. Let L be a pretransitive logic, $h, k < \omega$. If L[h] is k-tabular, then for all k-formulas φ we have

$$\mathrm{L}[h + 1] \vdash \varphi \text{ iff } \mathrm{L} \vdash \bigwedge_{i \leq h} (\Box^*(\Box^*\varphi \to \mathbf{B}_{i,k}) \to \mathbf{B}_{i,k}). \tag{17}$$

Proof. We may assume that both L and L[h+1] are consistent (Proposition 7). Let F be the k-canonical frame of L.

Suppose $\mathrm{L}[h + 1] \vdash \varphi$. We claim that for all $i \leq h$, $\Box^*(\Box^*\varphi \to \mathbf{B}_{i,k}) \to \mathbf{B}_{i,k}$ is true at every point x in the k-canonical model of L. Let $\neg\mathbf{B}_{i,k}$ be in x. Let us show that $\neg\mathbf{B}_{i,k} \wedge \Box^*\varphi \in y$ for some y with xR_{F}^*y. First, assume that x is in F[h+1]. By Proposition 8, x contains L[h+1]. Since $\varphi \in \mathrm{L}[h+1]$, we have $\Box^*\varphi \in \mathrm{L}[h+1]$. Thus

2343

$\Box^* \varphi \in x$. Since R_F^* is reflexive, in this case we can put $y = x$. Second, suppose x is not in $F[h+1]$. By Theorem 5, there exists y such that xR_F^*y and $\mathrm{ht}(F\langle y\rangle) = h+1$. We have $\Box^* \varphi \in y$ and $\mathbf{B}_{i,k} \notin y$. This proves the "only if" part.

Now suppose that $L \vdash \Box^*(\Box^* \varphi \to \mathbf{B}_{i,k}) \to \mathbf{B}_{i,k}$ for all $i \leq h$. Assume $\mathrm{ht}(F\langle x\rangle) = i \leq h+1$. In this case $\mathbf{B}_{i-1,k} \notin x$. Since $\Box^*(\Box^* \varphi \to \mathbf{B}_{i-1,k}) \to \mathbf{B}_{i-1,k}$ is in x, it follows that $\neg\mathbf{B}_{i-1,k} \wedge \Box^* \varphi$ is in y for some y with xR_F^*y. The first conjunct says that y is not in $F[i-1]$. Since y is in $F[i]$, it follows that $\mathrm{ht}(F\langle y\rangle) = i$. Hence y and x belong to the same cluster. Since $\Box^* \varphi \in y$, we obtain $\varphi \in x$. It follows that $\varphi \in x$ for all x in $F[h+1]$. By Proposition 8, $L[h+1] \vdash \varphi$. $\qquad\square$

Note that $\mathbf{B}_{0,k}$ is $\bot$ for all $k < \omega$. Thus, (17) generalizes the translation described in Theorem 4.

Theorem 6 provides translations for the case when L is a unimodal transitive logic (recall that transitive logics of finite height are locally tabular [22]). It should be noted that in this case Theorem 5, the key ingredient of the proof of Theorem 6, has been known since 1970s: formulas $\mathbf{B}_{i,k}$ in transitive canonical frames were described in [24] (see also [9], [1]).

An analog of Theorem 6 can be formulated for intermediate logics. Formulas $\mathbf{B}_{i,k}^I$ defining points of finite depth in finitely generated intuitionistic canonical frames were described in [25] (see also [2]). Similarly to the proof of Theorem 6, it can be shown that

$$IL[h+1] \vdash \varphi \text{ iff } IL \vdash \bigwedge_{i \leq h} ((\varphi \to \mathbf{B}_{i,k}^I) \to \mathbf{B}_{i,k}^I)$$

for all finite h and for all k-formulas φ.

5 Corollaries, examples, and open problems

The translation (17) holds for all finite h, k in the case when L is a transitive unimodal logic. Indeed, by the Segerberg – Maksimova criterion (Theorem 1), a transitive logic is locally tabular iff it is of finite height. This criterion was recently generalized to a wide family of pretransitive logics [23]. For example, if a unimodal L contains the formula $\Diamond^{m+1}p \to \Diamond p \vee p$ for some $m > 0$, then L is locally tabular iff it is of finite height. Thus, (17) holds for all finite h, k in this case too.

However, in general k-tabularity of $L[h]$ depends both on h and on k.

Example 3. Consider the smallest reflexive 2-transitive logic $K + \{p \to \Diamond p, \Diamond^3 p \to \Diamond^2 p\}$ and its extension L with the McKinsey formula for the master modality,

$\Box^2 \Diamond^2 p \to \Diamond^2 \Box^2 p$. Maximal clusters in the canonical frames of L are reflexive singletons, so $L[1] = K + p \leftrightarrow \Box p$ by Proposition 8. Clearly, $L[1]$ is locally tabular. It follows that we have the translation (17) from $L[2]$ to L for all finite k.

However, $L[2]$ is not even 1-tabular. To see this, consider the frame $F_0 = (\omega, R_0)$, where

$$x R_0 y \quad \text{iff} \quad x \neq y + 1 \text{ and } y \neq x + 1.$$

Let $F = (\omega + 1, R)$, where $x R y$ iff $x R_0 y$ or $y = \omega$. Clearly, $F \vDash L[2]$. Consider a model M on F such that $x \vDash p_0$ iff $x = 0$ or $x = \omega$. Put $\alpha_0 = p_0 \wedge \Diamond \neg p_0$, $\alpha_1 = \neg \Diamond \alpha_0 \wedge \neg p_0$, and $\alpha_{i+1} = \neg(\Diamond \alpha_i \vee \alpha_{i-1}) \wedge \neg p_0$ for $i > 0$. By an easy induction, in M we have for all i: $x \vDash \alpha_i$ iff $x = i$. Thus if $i \neq j$, then $\alpha_i \leftrightarrow \alpha_j \notin L$.

It is not difficult to construct other examples of this kind for arbitrary finite h: there are pretransitive logics such that $L[h]$ is locally tabular, and $L[h + 1]$ is not one-tabular.

With the parameter k, the situation is much more intriguing. The following result was proved in [18]:

$$\text{A unimodal transitive logic is locally tabular iff it is 1-tabular.} \quad (18)$$

The recent results [23] show that this equivalence also holds for many non-transitive logics. For example, if a unimodal L contains $\Diamond^{m+1} p \to \Diamond p \vee p$ for some $m > 0$, then it is locally tabular iff it is 1-tabular. The question whether this equivalence holds for every modal logic has been open since 1970s.

Theorem 7. There exists a unimodal 1-tabular logic L which is not locally tabular.

Proof (sketch). Let L be the logic of the frame $(\omega + 1, R)$, where

$$x R y \text{ iff } x \leq y \text{ or } x = \omega.$$

First, we claim that L is not locally tabular.

The following fact follows from Theorem 4.3 and Lemma 5.9 in [23]: if the logic of a frame F is locally tabular, then the logic of an arbitrary restriction $F{\restriction}V$ of F is locally tabular.

The restriction of $(\omega + 1, R)$ onto ω is the frame $(\omega, \leq)$, whose logic is not locally tabular (it is of infinite height). Thus L is not locally tabular.

To show that L is 1-tabular, we need the following observation. If every k-generated subalgebra of an algebra A contains at most m elements for some fixed $m < \omega$, then the free k-generated algebra in the variety generated by A is finite; see [19].

Consider the complex algebra A of the frame $(\omega+1, R)$. One can check that every 1-generated subalgebra of A contains at most 8 elements. By the above observation, L is 1-tabular. □

It is unknown whether 2-tabularity of a modal logic implies its local tabularity. At least, does k-tabularity imply local tabularity, for some fixed k for all unimodal logics? The same questions are open in the intuitionistic case [5, Problem 2.4].

Finite height is not a necessary condition for local tabularity of intermediate logics. What can be an analog of Gliveko's translation in the case of a locally tabular intermediate logic with no finite height axioms? Another generalization can probably be found in the area of modal intuitionistic logics. In [3], Glivenko type theorems were proved for extensions of the logic MIPC; in [4], local tabularity of these extensions was considered. What can be an analog of Theorem 6 for modal intuitionistic logics?

In [21], Glivenko's theorem was used to obtain decidability (and the finite model property) for extensions of S4 with $\Box\Diamond$-formulas (such formulas are built from literals $\Box\Diamond p_i$). An analog of this result can be obtained for extensions of a pretransitive logic L in the case when L is decidable (or has the finite model property) and L[1] is locally tabular.

References

[1] F. Bellissima. An effective representation for finitely generated free interior algebras. *Algebra Universalis*, 20(3):302–317, Oct 1985.

[2] F. Bellissima. Finitely generated free Heyting algebras. *Journal of Symbolic Logic*, 51(1):152–165, 1986.

[3] G. Bezhanishvili. Glivenko type theorems for intuitionistic modal logics. *Studia Logica*, 67(1):89–109, 2001.

[4] G. Bezhanishvili and R. Grigolia. Locally tabular extensions of MIPC. In *Advances in Modal Logic*, pages 101–120, 1998.

[5] G. Bezhanishvili and R. Grigolia. Locally finite varieties of heyting algebras. *Algebra universalis*, 54:465–473, 2005.

[6] P. Blackburn, M. de Rijke, and Y. Venema. *Modal logic*. Cambridge University Press, 2002.

[7] M. Byrd. On the addition of weakened L-reduction axioms to the Brouwer system. *Mathematical Logic Quarterly*, 24(25-30):405–408, 1978.

[8] A. Chagrov and M. Zakharyaschev. *Modal logic*, volume 35 of *Oxford Logic Guides*. Oxford University Press, 1997.

[9] K. Fine. Logics containing K4. Part II. *Journal of Symbolic Logic*, 50(3):619–651, 09 1985.

[10] G. Fischer-Servi. Axiomatizations for some intuitionistic modal logics. *Rendiconti di Matematica di Torino*, 42:179–194, 1984.

[11] D. Gabbay, A. Kurucz, F. Wolter, and M. Zakharyaschev. *Many-dimensional modal logics: theory and applications*. Studies in Logic and the Foundations of Mathematics. Elsevier, 2003.

[12] D. Gabbay, V. Shehtman, and D. Skvortsov. *Quantification in nonclassical logic*. Elsevier, 2009.

[13] V. Glivenko. Sur quelques points de la logique de M. Brouwer. *Bulletins de la classe des sciences*, 15:183–188, 1929.

[14] R. Goldblatt. *Mathematics of modality*. CSLI Publications, 1993.

[15] M. Kracht. *Tools and techniques in modal logic*. Elsevier, 1999.

[16] A. Kudinov and I. Shapirovsky. Partitioning Kripke frames of finite height. *Izvestiya: Mathematics*, 81(3):592, 2017.

[17] D. Makinson. Non-equivalent formulae in one variable in a strong omnitemporal modal logic. *Math. Log. Q.*, 27(7):111–112, 1981.

[18] L. Maksimova. Modal logics of finite slices. *Algebra and Logic*, 14(3):304–319, 1975.

[19] A. Malcev. *Algebraic systems*. Springer, 1973.

[20] K. Matsumoto. Reduction theorem in Lewis's sentential calculi. *Mathematica Japonicae*, 3:133–135, 1955.

[21] V. Rybakov. A modal analog for glivenko's theorem and its applications. *Notre Dame J. Formal Logic*, 33(2):244–248, 03 1992.

[22] K. Segerberg. *An essay in classical modal logic*. Filosofska Studier, vol.13. Uppsala Universitet, 1971.

[23] I. Shapirovsky and V. Shehtman. Local tabularity without transitivity. In *Advances in Modal Logic*, volume 11, pages 520–534. College Publications, 2016.

[24] V. Shehtman. Rieger-nishimura lattices. *Doklady Mathematics*, 19:1014–1018, 1978.

[25] V. Shehtman. *Applying Kripke models to the investigation of superintuitionistic and modal logics*. PhD thesis, Moscow State University, 1983. In Russian.

[26] A. Simpson. *The proof theory and semantics of intuitionistic modal logic*. PhD thesis, University of Edinburgh, 1994.

Received 21 October 2017

A Modal Characterisation of an Intuitionistic I/O Operation

Xavier Parent*
TU Wien
xavier@logic.at

Abstract

It is known that Makinson and van der Torre's basic I/O operation out_2 can faithfully be "embedded" into (or "encoded" in) classical modal logic. It is shown that an analogous result holds for the intuitionistic variant of out_2. The target of the embedding is the constructive modal logic CK that evolved through work of Wijesekera, Mendler, de Paiva and Ritter. The very same translation that embeds out_2 into classical modal logic is used.

1   Introduction

Due to Makinson and van der Torre [14, 15], input/output (I/O) logic aims at generalizing the theory of conditional obligation from modal logic [12, 13] to the abstract study of conditional codes viewed as relations between classical formulae. The meaning of the normative concepts is given in terms of a set of procedures yielding outputs for inputs. Detachment (or modus ponens) is the core mechanism of the semantics being used. A number of I/O operations are studied in the aforementioned paper [14]. It is shown that they correspond to a series of proof systems of increasing strength. I/O logic belongs to the category of what has been called "norm-based semantics" by Hansen [11, p. 288]. The core idea is to explain the principles of deontic logic, not by some set of possible worlds among which some are ideal or at least better than others, but with reference to a set of explicit norms or existing standards. There are at least two reasons for the recent growth of interest in this approach. First, such a semantics allows one to remain neutral on a number of controversial issues, like the question of whether norms bear truth-values [14],

---

*I wish to thank two anonymous referees for valuable comments that helped improve the paper. Discussions with Dov Gabbay were helpful. Feedback received at IMLA 2017 was also useful.

or the question of whether normative statements are based on a maximization process [22]. Second, the norm-based approach has proven to be a fruitful addition to our understanding of key issues in deontic reasoning, like the question of how to model permissions [16, 5, 29], the issue of how to accommodate and resolve conflicts between norms [19], and the question of how to reason about norm violation [15]. As is well-known, these issues highlighted limitations of so-called Standard Deontic Logic (SDL) and its Kripke-type possible worlds semantics, with which philosophers may be more familiar (see, e. g., [24]).

These developments will not be discussed in this paper. For an overview, see [21]. Here I will not go beyond the basic set-up used by Makinson and van der Torre in their [14], except for the following. They use classical propositional logic as the base logic. Parent & al. [20] study the effects of using intuitionistic propositional logic (IPL). It is shown that three of the four standard, classically-based I/O operations have an axiomatizable intuitionistic version. These are: the simple-minded I/O operation out_1; the basic I/O operation out_2; and the reusable I/O operation out_3. Of these, the most striking one is undoubtedly out_2. I will be primarily concerned with it. From now onward I will refer to this one as out_2^i, where the superscript i is mnemonic for "intuitionistic". The basic idea is to replace in the semantic idiom the notion of maximal consistent set by its intuitionistic counterpart, the notion of saturated set. The main observation made in [20] is that one obtains the same syntactic characterization of the input/output system, up to the meaning of the connectives. This observation is shown to carry over to the intuitionistic versions of out_1 and out_3. The question of whether it also applies to that of out_4 is left unanswered.

This paper will address another issue left open in [20]. Makinson and van der Torre [14] show that their I/O operations out_2 and out_4 can be reformulated in terms of modal logic. The essential idea is to prefix heads of rules with boxes and apply a suitable modal logic. It is natural to ask if an analog result holds in an intuitionistic setting. The answer to this question turns out to be positive, at least for out_2^i. Admittedly this is a small point, but one (I believe) that is worth clarifying. The intuitionistic modal logic into which out_2^i will be embedded is the system called CK (for constructive K) by Mendler and de Paiva [17] and de Paiva and Ritter [8]. CK is much like (the propositional fragment of) a prior system by Wijesekera [31]. They share the feature that $\Diamond$ does not distribute over disjunction. But CK also rejects the nullary version of the law of distributivity, $\neg \Diamond \bot$, *aka* $\Diamond \bot \rightarrow \bot$. On the semantic side, this is made possible by allowing non-normal (or, as de Paiva and colleagues call them, "fallible" or inconsistent) worlds in the models.

The main result in the paper is a faithful embedding theorem, which echoes the one established by Makinson and van der Torre in the original setting. The theorem is proved for the $\Diamond$-free, first-degree fragment of CK–that is, the subsystem of CK in which formulas do not contain occurrences of $\diamond$ and nested occurrences of $\square$. But I will present the full system in order to make the paper self contained.

Such a result is interesting in its own right, because it makes a bridge between two independent frameworks. This bridge can be used to import results, ideas, and techniques from one to the other. For instance, it can unlock the door to an automation of the source logic. Benzmüller & al. [3] implement the standard I/O operations out_2 and out_4 in Isabelle/HOL [18] via an implementation of their modal translation, making use of the so-called shallow semantical embedding of modal systems K and T into HOL [4]. The embeddings are encoded in Isabelle/HOL for automation.

The layout of this paper is as follows. Section 2 provides the reader with the required background. Section 3 describes the embedding into CK. Section 4 ends with a number of open issues.

2 Background

I start by explaining the basic idea underpinning the I/O framework. In I/O logic, a conditional obligation is represented as a pair (a, x) of propositional formulas, where a is the body (antecedent) and x is the head (consequent). Intuitively, (a, x) may be read as "if a is the case, then x is obligatory". A normative system N is a set of such pairs. Let A be a set of formulas. The main construct has the form: $x \in out(N, A)$. Intuitively this can be read as follows: given input set A (state of affairs), x (obligation) is outputted under norms N.

2.1 Intuitionistic Basic I/O Operation

This section describes the intuitionistic variant of the basic I/O operation out_2 initially put forth by Makinson and van der Torre [14]. The operation is denoted by out_2^i, where the superscript i stands for "intuitionistic". This material is taken from [20].

Throughout this paper, $\mathcal{L}_{\mathsf{IPL}}$ is the set of all formulas in the language of intuitionistic propositional logic. I use the system put forth by Thomason [30]. $\vdash_{\mathsf{IPL}}$ is the derivability relation in this logic. Cn_{IPL} denotes the associated consequence operation, viz. $Cn_{\mathsf{IPL}}(S) = \{a : S \vdash_{\mathsf{IPL}} a\}$, where S is a set of formulas in $\mathcal{L}_{\mathsf{IPL}}$. A set S of formulas is said to be consistent in IPL if there is a wff a such that $S \not\vdash_{\mathsf{IPL}} a$.

Definition 1 (Saturated set, [30]). *Let S be a non-empty set of formulas in $\mathcal{L}_{\mathsf{IPL}}$.*

S is said to be saturated if the following three conditions hold:

$$S \text{ is consistent in IPL} \tag{1}$$

$$a \vee b \in S \Rightarrow a \in S \text{ or } b \in S \text{ (S is join-prime)} \tag{2}$$

$$S \vdash_{\mathsf{IPL}} a \Rightarrow a \in S \text{ (S is closed under } \vdash_{\mathsf{IPL}}) \tag{3}$$

Definition 2 implements the notion of (single-step) detachment or modus ponens. It is the *modus operandi* of the semantics.

Definition 2 (Image). *Let A be a set of formulas. $N(A) = \{x : (a, x) \in N$ for some $a \in A\}$. For $N(A)$, read "the N of A".*

Intuitively, $N(A)$ gathers the heads of all the conditional obligations (a, x) in N that are "triggered" by set A. As argued by Boghossian [6], detachment is part of the meaning of a conditional statement. Hence the idea of making detachment the core mechanism of the semantics.[1]

Definition 3 (out_2^i, intuitionistic basic output).

$$out_2^i(N, A) = \begin{cases} \cap\{Cn_{\mathsf{IPL}}(N(S)) : A \subseteq S, S \text{ saturated}\}, & \text{if } A \text{ is consistent in IPL} \\ Cn_{\mathsf{IPL}}(h(N)), & \text{otherwise} \end{cases}$$

where $h(N)$ is the set of all heads of elements of N, viz. $h(N) = \{x : (a, x) \in N$ for some $a\}$.

Our first observation follows at once from Definition 3 and the property of monotony of $\vdash_{\mathsf{IPL}}$. This property tells us that $\Gamma \vdash_{\mathsf{IPL}} x$ whenever $\Delta \vdash_{\mathsf{IPL}} x$ and $\Delta \subseteq \Gamma$.

Fact 4. $out_2^i(N, A) \subseteq Cn_{\mathsf{IPL}}(h(N))$.

Put $out_2^i(N) = \{(A, x) : x \in out_2^i(N, A)\}$. This definition leads to an axiomatic characterization that is much like those used for conditional logic. The specific rules of interest here are described below. They are formulated for a singleton input set A (for such an input set, curly brackets will be omitted). The move to an input set A of arbitrary cardinality will be explained in a moment.

---

[1]Such a motivation is not in the original papers [14, 15]. It is given and discussed in more detail in [22].

$$\text{SI } \frac{(a, x) \qquad b \vdash_{\mathsf{IPL}} a}{(b, x)} \qquad\qquad \text{WO } \frac{(a, x) \qquad x \vdash_{\mathsf{IPL}} y}{(a, y)}$$

$$\text{AND } \frac{(a, x) \qquad (a, y)}{(a, x \wedge y)} \qquad\qquad \text{OR } \frac{(a, x) \qquad (b, x)}{(a \vee b, x)}$$

SI and WO abbreviate "strengthening of the input" and "weakening of the output", respectively. IPL is known to be decidable, and thus the relation expressed by each rule is decidable, as is usually required for the rules of an axiom system.

Given a set of rules, a derivation from a set N of pairs (a, x) is a sequence $\alpha_1, ..,$ α_n of pairs of formulas such that for each index $0 \leq i \leq n$ one of the following holds:

- α_i is an hypothesis, i.e. $\alpha_i \in N$;

- α_i is $(\top, \top)$, where $\top$ is a zero-place connective standing for 'tautology';

- α_i is obtained from preceding element(s) in the sequence using one of $\{\text{SI, WO,}$ AND, OR$\}$.

All elements in the sequence are pairs of the form (a, x). Derivation steps done in the base logic IPL are not part of it.

A pair (a, x) of formulas is said to be derivable from N if there is a derivation from N whose final term is (a, x). This will be written as $(a, x) \in deriv_2^i(N)$.

When A is a set of formulas, derivability of (A, x) from N is defined as derivability of (a, x) from N for some conjunction $a = a_1 \wedge ... \wedge a_n$ of elements of A. I understand the conjunction of zero formulas to be a tautology, so that $(\emptyset, a)$ is derivable from N if and only if (iff) $(\top, a)$ is.

Let $deriv_2^i(N, A) = \{x : (A, x) \in deriv_2^i(N)\}$. We have:

Theorem 5 (Soundness and completeness). $out_2^i(N, A) = deriv_2^i(N, A)$

Proof. This is [20, Theorem 13]. $\qquad\qquad\qquad\qquad\qquad\qquad\qquad\qquad\qquad$ $\square$

2.2 Constructive Modal Logic CK

This section describes the system of constructive modal logic called CK (for constructive K) by Mendler and de Paiva [17] and de Paiva and Ritter [8].

The language is denoted by $\mathcal{L}_{\mathsf{CK}}$. It is obtained by adding to the language of IPL the two modal operators $\square$ and $\diamond$. For simplicity's sake, $\bot$ is identified with a privileged atomic sentence, as in so-called minimal logic.

Definition 6. *A Kripke model of* CK *is a structure* $M = (W, \leq, R, v)$, *where* W *is a non-empty set of possible worlds (or points),* $\leq$ *is a reflexive and transitive binary relation on* W, R *is a binary relation on* W, *and* v *is a function assigning to each propositional letter* p *a subset of* W, *viz* $v(p) \subseteq W$. *Furthermore,* $\leq$ *is required to be hereditary with respect to propositional variables:*

$$\text{If } w \leq w' \text{ and } w \in v(p), \text{ then } w' \in v(p)$$

$\leq$ is used to express the forcing condition for the arrow connective $\rightarrow$, whilst R (with a little help from $\leq$) is employed to articulate the forcing condition for the modal operators $\square$ and $\diamond$.

Definition 7 (Forcing). *Given a model* $M = (W, \leq, R, v)$, *and a world* $w \in W$, *the forcing relation* $M, w \vDash a$ *(read as "formula* a *is 'forced' at world* w *in model* M*") is defined by induction on the structure of formula* a *using the following clauses:*

- $M, w \vDash p$ *iff* $w \in v(p)$

- $M, w \vDash \top$

- $M, w \vDash b \wedge c$ *iff* $M, w \vDash b$ *and* $M, w \vDash c$

- $M, w \vDash b \vee c$ *iff* $M, w \vDash b$ *or* $M, w \vDash c$

- $M, w \vDash b \rightarrow c$ *iff* $(\forall w')\, (w \leq w' \Rightarrow (M, w' \vDash b \Rightarrow M, w' \vDash c))$

- $M, w \vDash \square b$ *iff* $(\forall w')\, (w \leq w' \Rightarrow \forall u\, (w'Ru \Rightarrow M, u \vDash b))$

- $M, w \vDash \diamond b$ *iff* $(\forall w')\, (w \leq w' \Rightarrow \exists u\, (w'Ru\ \&\ M, u \vDash b))$

As usual I will drop reference to M, and write $w \vDash a$, when it is clear what model is intended.

A world w is said to be normal if $w \nvDash \bot$, and non-normal (or fallible) if $w \vDash \bot$. The following two constraints are placed on models:

If w is non-normal and $w \leq w'$ or wRw', then w' is non-normal $\qquad$ (c_1)

If w is non-normal, then $M, w \vDash p$ for all propositional letters p $\qquad$ (c_2)

(c_1) and (c_2) imply that, for all formula a, $M, w \vDash a$, whenever w is non-normal.

Following Fitting [10], Mendler and de Paiva [17] introduce a "hybrid" notion of consequence, which distinguishes between global and local assumptions. Global (or universal) assumptions are required to hold at all points in a given model, while local assumptions are required to hold at a given point in that model. I will use a local consequence relation instead. A formula a is said to be a semantic consequence of A (notation: $A \models a$), whenever, for every model M, and for all worlds w in M, if all of A hold at w, then so does a. My reason for doing so is twofold. First, it will simplify the arguments. Second, the contrast between global and local assumptions will not play any role in subsequent developments.

CK comes with a Hilbert-style proof system, whose axioms consist of all the validities of the intuitionistic propositional logic IPL together with

$$\Box(a \rightarrow b) \rightarrow (\Box a \rightarrow \Box b) \tag{K-$\Box$}$$

$$\Box(a \rightarrow b) \rightarrow (\Diamond a \rightarrow \Diamond b) \tag{K-$\Diamond$}$$

CK also has the rule of modus ponens and the rule of necessitation for $\Box$. As usual, $\vdash_{\mathsf{CK}} a$ indicates that a is a theorem in CK, and $A \vdash_{\mathsf{CK}} a$ indicates that the formula a is in CK a deductive consequence of the set of (local) assumptions A. We have $A \vdash_{\mathsf{CK}} a$ whenever there is a finite $A' \subseteq A$ such that $\vdash_{\mathsf{CK}} \bigwedge A' \rightarrow a$. The limiting case where $\bigwedge \emptyset = \top$ is included.

The soundness and completeness theorem is stated below.

Theorem 8. $A \models a$ iff $A \vdash_{\mathsf{CK}} a$.

Proof. This is Mendler and de Paiva [17, Theorem 1]. $\qquad\qquad\qquad\Box$

3 Modal Embedding Result

The intuitionistic analog of Lindenbaum's lemma will be needed. It reads:

Lemma 9. *Let $A \cup \{a\} \subseteq \mathcal{L}_{\mathsf{IPL}}$. If $A \nvdash_{\mathsf{IPL}} a$, then there is a saturated set S of formulas (in $\mathcal{L}_{\mathsf{IPL}}$) such that $A \subseteq S$ and $a \notin S$.*

Proof. This is [30, Lemma 1]. $\qquad\qquad\qquad\qquad\qquad\qquad\qquad\qquad\qquad\Box$

The following observation will also come in handy.

Theorem 10. *Let A be a non-empty set of formulas in $\mathcal{L}_{\mathsf{IPL}}$. A is consistent in CK if and only if A is consistent in IPL.*

Proof. For the left-to-right direction, suppose A is consistent in CK. By Theorem 8, A is satisfiable in a model $M = (W, \leq, R, v)$ of CK. That is, there is a normal world w in M such that $w \models x$ for all $x \in A$. Let $M^w = (W^w, \leq^w, v^w)$, where

- $W^w = \{u \in W : u$ is normal $\& \ w \leq u\}$

- $\leq^w = \leq \cap (W^w \times W^w)$

- $v^w(p) = v(p) \cap W^w$ for all propositional letters p

M^w is an ordinary Kripke model of IPL. An easy induction establishes that each world in M^w forces the same formulas $a \in \mathcal{L}_{\mathsf{IPL}}$ as in M. Hence, A is satisfiable in an ordinary Kripke model of IPL. By soundness, A is consistent in IPL.

The proof of the right-to-left direction is similar. Starting with a model M of IPL in which A is satisfiable, one needs to get a model M' of CK in which A is also satisfiable. M' shares W, $\leq$ and v with M. Its new component R is the identity relation. In M', constraints (c_1) and (c_2) are trivially verified, because all the worlds are normal. $\qquad\square$

The very same translation that embeds the original I/O logic into classical modal logic is used. The core idea is to convert each pair in N into an intuitionistic implication whose head is prefixed with $\square$, and then use CK to calculate the output. The main result in this paper is that such an embedding is faithful. The exact statement of the result to be established is given by equation (4) where $N^\square = \{a \to \square x : (a, x) \in N\}$:

$$x \in \mathit{deriv}_2^i(N, A) \Leftrightarrow h(N) \vdash_{\mathsf{IPL}} x \text{ and } N^\square \cup A \vdash_{\mathsf{CK}} \square x \qquad (4)$$

The left-to-right (LTR) implication says that the translation "preserves" derivability of outputs, while the right-to-left (RTL) implication says that no new outputs can be derived. Below each direction is established in turn.

Theorem 11 (Faithfulness, LTR). *If $x \in \mathit{deriv}_2^i(N, A)$, then $h(N) \vdash_{\mathsf{IPL}} x$ and $N^\square \cup A \vdash_{\mathsf{CK}} \square x$.*

Proof. Assume $x \in \mathit{deriv}_2^i(N, A)$. The claim $h(N) \vdash_{\mathsf{IPL}} x$ follows from Theorem 5 and Fact 4.

By definition of deriv_2^i, $(a, x) \in \mathit{deriv}_2^i(N)$, for a conjunction $a = a_1 \wedge \dots \wedge a_n$ of elements in A. One shows that $N^\square \cup \{a\} \vdash_{\mathsf{CK}} \square x$ by a straightforward induction on the length of the derivation of (a, x):

Base case: (a, x) has a derivation of length 1. In that case, either (a, x) is $(\top, \top)$ or $(a, x) \in N$. The claim $N^\square \cup \{a\} \vdash_{\mathsf{CK}} \square x$ holds, because each of $\top \to \square\top$ and $((a \to \square x) \wedge a) \to \square x$ is a theorem in CK;

Inductive step: (a, x) has a derivation of length $n+1$. The interesting case is when (a, x) is obtained from earlier lines by a derivation rule. Only two $\square$-principles are needed. One is the axiom K-$\square$. It is needed to handle WO. The other is $(\square a \wedge \square b) \to \square(a \wedge b)$. It is needed to handle AND, and is derivable in CK.

The claim $N^{\square} \cup A \vdash_{\mathsf{CK}} \square x$ follows from $N^{\square} \cup \{a\} \vdash_{\mathsf{CK}} \square x$ and the principle of cumulative transitivity for $\vdash_{\mathsf{CK}}$. This principle tells us that $\Delta \cup \Gamma \vdash_{\mathsf{CK}} y$ whenever $\Gamma \vdash_{\mathsf{CK}} b$ and $\Delta \cup \{b\} \vdash_{\mathsf{CK}} y$. $\qquad\square$

Theorem 12 (Faithfulness, RTL). *If both $h(N) \models_{\mathsf{IPL}} x$ and $N^{\square} \cup A \vdash_{\mathsf{CK}} \square x$, then $x \in deriv_2^i(N, A)$.*

Proof. I show the contrapositive. Assume $x \notin deriv_2^i(N, A)$ and $h(N) \vdash_{\mathsf{IPL}} x$. To show: $N^{\square} \cup A \nvdash_{\mathsf{CK}} \square x$. Our aim is to establish that $N^{\square} \cup A \not\models \square x$. The desired conclusion, $N^{\square} \cup A \nvdash_{\mathsf{CK}} \square x$, follows at once from this and the soundness half of Theorem 8.

By Theorem 5, $x \notin out_2^i(N, A)$. So $out_2^i(N, A) \neq Cn_{\mathsf{IPL}}(h(N))$. By Definition 3, A is consistent in IPL and $out_2^i(N, A) = \cap\{Cn_{\mathsf{IPL}}(N(S)) : A \subseteq S, S \text{ saturated}\}$. So, since $x \notin out_2^i(N, A)$, there is some saturated set $S \supseteq A$ with $x \notin Cn_{\mathsf{IPL}}(N(S))$. Define $M = (W, \leq, R, v)$ as follows:

- $W = \{w : w \text{ is a saturated set of formulas in } \mathcal{L}_{\mathsf{IPL}}\}$

- $w \leq u$ iff $w \subseteq u$

- wRu iff: for all $(b, y) \in N$, if $b \in w$, then $y \in u$

- $v(p) = \{w : p \in w\}$

M is a model of CK. By construction, $S \in W$. The following observation will come in handy.

Claim 13. *Let x be a formula in $\mathcal{L}_{\mathsf{IPL}}$. For all $w \in W$, $x \in w$ iff $M, w \models x$.*

Proof of Claim 13. By induction on x. I consider only the case where x is a conditional, $b \to c$, focusing on the proof of the right-to-left direction. Assume $b \to c \notin w$. Since a saturated set is closed under $\vdash_{\mathsf{IPL}}$, Definition 1, $w \nvdash_{\mathsf{IPL}} b \to c$. By the deduction theorem, $w \cup \{b\} \nvdash_{\mathsf{IPL}} c$. By Lemma 9, there is a saturated set u such that $w \cup \{b\} \subseteq u$ and $c \notin u$. On the one hand, $w \leq u$. On the other hand, the inductive hypothesis yields $u \models b$ and $u \not\models c$, which suffices for $w \not\models b \to c$. $\qquad\square$

Claim 14 below will help us establish the desired intermediate conclusion, viz. $N^{\square} \cup A \not\models \square x$.

Claim 14. *The following holds in M:*

$$\text{For all } a \in A, S \models a \tag{5}$$
$$\text{For all } b \to \square y \in N^{\square}, S \models b \to \square y \tag{6}$$
$$S \not\models \square x \tag{7}$$

Proof of Claim 14. (5) follows easily from Claim 13 and $A \subseteq S$. For (6), let $b \to \Box y \in N^\Box$. Let t be such that $S \leq t$ and $t \models b$. Let u and v be such that $t \leq u$ and uRv. The formula b is in $\mathcal{L}_{\mathsf{IPL}}$. By Claim 13, $b \in t \subseteq u$. Since $(b, y) \in N$ and uRv, $y \in v$. By Claim 13 again, $v \models y$, since y is in $\mathcal{L}_{\mathsf{IPL}}$ too. By the forcing condition for $\Box$, $t \models \Box y$. By the forcing condition for $\to$, $S \models b \to \Box y$. Hence, for all $b \to \Box y \in N^\Box$, $S \models b \to \Box y$.

For (7), recall that $N(S) \nvdash_{\mathsf{IPL}} x$. By Lemma 9, there is a saturated set t such that $N(S) \subseteq t$ and $x \notin t$. On the one hand, $t \in W$. On the other hand, x is a formula in $\mathcal{L}_{\mathsf{IPL}}$. So $t \nvDash x$, by Claim 13. Let $(b, y) \in N$. Suppose $b \in S$. By construction, $y \in N(S) \subseteq t$. Hence, $y \in t$, which suffices for SRt. Trivially $S \leq S$. By the forcing condition for $\Box$, $S \nvDash \Box x$ as required. $\qquad\square$

This concludes the proof of Theorem 12. $\qquad\square$

It is worthwhile to mention that the proofs of Theorems 11 and 12 also go through in Wijesekera's initial system. Thus, the proposed embedding works in both systems. However, the proof of Theorem 12 does not carry over to the constructive modal logic $\mathsf{CS4}$ (see, e.g., [1, 8]). $\mathsf{CS4}$ is obtained by supplementing CK with the T-axioms $\Box x \to x$, $x \to \Diamond x$ as well as the S4-axioms $\Box x \to \Box\Box x$, $\Diamond x \to \Diamond\Diamond x$. It is characterized by the class of models in which R is in addition reflexive and transitive, and R and $\leq$ are such that $(R\circ\leq) \subseteq (\leq\circ R)$ where $\circ$ denotes composition of relations. In the model M used in the proof of Theorem 12, the latter constraint is satisfied. But there is no guarantee that R is reflexive and transitive. Thus there is no guarantee that M is a model of $\mathsf{CS4}$.

One would like to know whether the embedding result extends to other systems in the so-called intuitionistic "modal cube" introduced in [28] or its constructive variant (See Figure 1). For a given system, call it S, to act as a substitute for CK, its $\Diamond$-free, first-degree fragment must coincide with that of CK. I would conjecture that this requirement is at least met for the systems between IK and $\mathsf{IK45}$ in the intuitionistic modal cube, and the systems between CK and $\mathsf{CK45}$ in its constructive variant. The detailed verification of this claim must be postponed until another occasion.

4 Conclusion

I conclude this paper by highlighting a number of issues to consider in future research besides the aforementioned one.

First, one would like to know if the embedding can be extended to the other intuitionistic I/O operations defined in [20]. The basic reusable I/O operation out_4^i

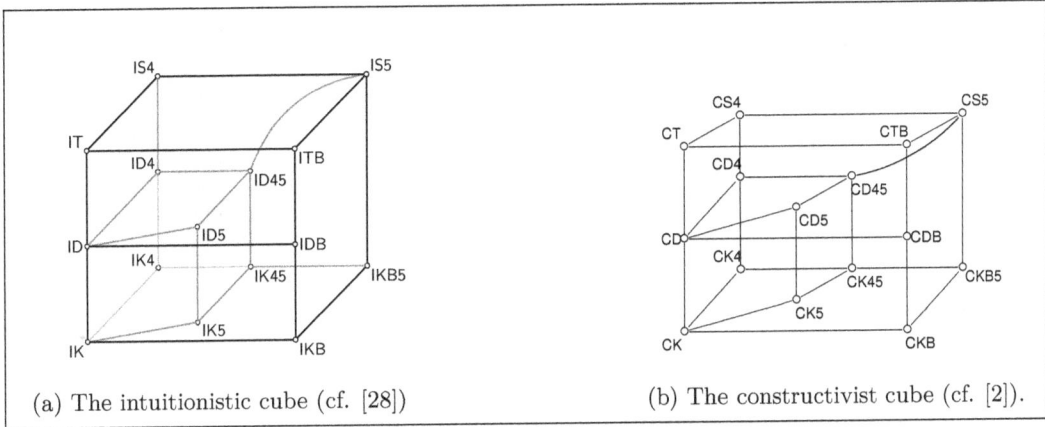

(a) The intuitionistic cube (cf. [28])

(b) The constructivist cube (cf. [2]).

Figure 1: The modal cubes.

is worth a mention. It is much like out_2^i, except that it also allows outputs to be recycled as inputs. On the syntactical side, we have in addition the rule of cumulative transitivity:

$$\text{CT} \; \frac{(a, x) \qquad (a \wedge x, y)}{(a, y)}$$

Makinson and van der Torre [14] show that the classically based out_4 can faithfully be embedded into a number of modal systems containing the T-axiom. It would be pleasant to be able to report that an analogous result holds for out_4^i, if one uses, e.g., the propositional fragment of Fitch's $\mathcal{M}$ [9] which is CK plus the T-axioms $\Box x \to x$, $x \to \Diamond x$. However, the fact that out_4^i still lacks an axiomatic characterization analogous to Theorem 5 presents a serious obstacle to obtaining such a result.

Second, I have confined myself to unconstrained I/O logic, which is usually considered just a stepping stone towards a finer-grained account of normative reasoning. The present account inevitably inherits the problems faced by unconstrained I/O logic, which have led to the further developments alluded to in the introductory section. In particular the present account puts aside the subtleties of contrary-to-duty (CTD) obligations. This can be illustrated with the "white fence" scenario due to Prakken and Sergot [27]: there should no fence; if there is a fence, it should be white; there is a fence. The encoding in CK gives: $N^\Box$ is $\{\top \to \Box\neg f, f \to \Box(w \wedge f)\}$ and A is $\{f\}$. One derives $\Box\bot$, which is the opposite of what we want. Drawing on analogous constructions in the logics of belief change and nonmonotonic inferences, the traditional approach in I/O logic consists in constraining the I/O operations to

2359

avoid output that is inconsistent with the input [15]. However, the systems of constrained I/O logic do not have a known axiomatic characterization. Furthermore, the (full join and meet) constrained I/O operations are in general nonmonotonic with respect to the input set A. It is unclear how they can be encoded in CK, whose consequence relation is monotonic. An alternative approach to CTDs has recently been studied in Parent and van der Torre [22, 23, 26, 25]. The unconstrained I/O operations are defined in such a way that they are not closed under the consequence relation of the base logic. Furthermore, some of these I/O operations have a built-in consistency check, which filters out excess output. This yields variant proof systems with neither the rule WO nor the zero-premise rule TAUT:[2]

$$\text{TAUT } \frac{-}{(\top, \top)}$$

The question remains open whether these variant systems have an intuitionistic counterpart that can be embedded into some existing (non-normal) constructive modal logic(s) or variant thereof [7].

References

[1] N. Alechina, M. Mendler, V. de Paiva, and E. Ritter. Categorical and Kripke semantics for constructive S4 modal logic. In L. Fribourg, editor, *Computer Science Logic: 15th International Workshop, CSL 2001, Proceedings*, pages 292–307, Berlin, Heidelberg, 2001. Springer.

[2] R. Arisaka, A. Das, and L. Straßburger. On nested sequents for constructive modal logics. *Logical Methods in Computer Science*, 11(3):1–33, 2015.

[3] C. Benzmüller, A. Farjami, P. Meder, and X. Parent. I/O logic in HOL. *IfColog Journal of Logics and their Applications (FLAP)*, 6(5):715–732, 2019.

[4] C. Benzmüller and L. Paulson. Multimodal and intuitionistic logics in simple type theory. *The Logic Journal of the IGPL*, 18(6):881–892, 2010.

[5] G. Boella and L. van der Torre. Institutions with a hierarchy of authorities in distributed dynamic environments. *Artificial Intelligence and Law*, 16(1):53–71, 2008.

[6] P. Boghossian. Knowledge of logic. In P. Boghossian and C. Peacocke, editors, *New Essays on the A Priori*, pages 229–254. Clarendon Press, Oxford, 2000.

[7] T. Dalmonte, C. Grellois, and N. Olivetti. Intuitionistic non-normal modal logics: A general framework. *Journal of Philosophical Logic*, 49:833–882, 2020.

[8] V. de Paiva and E. Ritter. Basic constructive modality. In J.-Y. Beziau and M. Coniglio, editors, *Logic without Frontiers–Festschrift for Walter Alexandre Carnielli on the Occasion of his 60th Birthday*, pages 411–428. College Publications, London, UK, 2011.

---

[2]TAUT is embedded in the definition of the notion of derivation given in Section 2.1.

[9] F. Fitch. Intuitionistic modal logic with quantifiers. *Portugaliae Mathematica*, 7(2):113–118, 1949.

[10] M. Fitting. Basic modal logic. In D. M. Gabbay, C. J. Hogger, and J. A. Robinson, editors, *Handbook of Logic in Artificial Intelligence and Logic Programming (Vol. 1)*, pages 368–448. Oxford University Press, New York, USA, 1993.

[11] J. Hansen. Reasoning about permission and obligation. In S. O. Hansson, editor, *David Makinson on Classical Methods for Non-Classical Problems*, pages 287–333. Springer, Dordrecht, 2014.

[12] B. Hansson. An analysis of some deontic logics. *Noûs*, 3:373–398, 1969.

[13] D. Makinson. Five faces of minimality. *Studia Logica*, 52(3):339–379, 1993.

[14] D. Makinson and L. van der Torre. Input/output logics. *Journal of Philosophical Logic*, 29(4):383–408, 2000.

[15] D. Makinson and L. van der Torre. Constraints for input/output logics. *Journal of Philosophical Logic*, 30(2):155–185, 2001.

[16] D. Makinson and L. van der Torre. Permissions from an input/output perspective. *Journal of Philosophical Logic*, 32(4):391–416, 2003.

[17] M. Mendler and V. de Paiva. Constructive CK for contexts. In L. Serafini and P. Bouquet, editors, *Proceedings Context Representation and Reasoning (CRR-2005)*, volume 136 of *CEUR*, 2005. Also presented at the Association for Symbolic Logic Annual Meeting, Stanford University, USA, 22nd March 2005.

[18] T. Nipkow, L. C. Paulson, and M. Wenzel. *Isabelle/HOL - A Proof Assistant for Higher-Order Logic*, volume 2283 of *Lecture Notes in Computer Science*. Springer, 2002.

[19] X. Parent. Moral particularism in the light of deontic logic. *Artificial Intelligence and Law*, 19(2-3):75–98, 2011.

[20] X. Parent, D. Gabbay, and L. van de Torre. Intuitionistic basis for input/output logic. In S. O. Hansson, editor, *David Makinson on Classical Methods for Non-Classical Problems*, pages 263–286. Springer, Dordrecht, 2014.

[21] X. Parent and L. van der Torre. Input/output logics. In D. Gabbay, J. Horty, X. Parent, R. van der Meyden, and L. van der Torre, editors, *Handbook of Deontic Logic and Normative Systems*, volume 1, pages 500–542. College Publications, London. UK, 2013.

[22] X. Parent and L. van der Torre. Aggregative deontic detachment for normative reasoning. In T. Eiter, C. Baral, and G. De Giacomo, editors, *Principles of Knowledge Representation and Reasoning. Proceedings of the Fourteenth International Conference (KR 2014)*, pages 646–649, Palo Alto, California, 2014. AAAI Press.

[23] X. Parent and L. van der Torre. "Sing and dance!"–Input/output logics without weakening. In Fabrizio Cariani, Davide Grossi, Joke Meheus, and Xavier Parent, editors, *Deontic Logic and Normative Systems–12th International Conference, DEON 2014, Ghent, Belgium, July 12-15, 2014. Proceedings*, volume 8554 of *Lecture Notes in Computer Science*, pages 149–165. Springer, 2014.

[24] X. Parent and L. van der Torre. *Introduction to Deontic Logic and Normative Systems*.

Texts in Logic and Reasoning. College Publications, London, UK, 2018.

[25] X. Parent and L. van der Torre. I/O logics with a consistency check. In J. Broersen, C. Condoravdi, N. Shyam, and G. Pigozzi, editors, *Deontic Logic and Normative Systems - 14th International Conference, DEON 2018, Utrecht, The Netherlands, July 3-6, 2018*, pages 285–299. College Publications, 2018.

[26] X. Parent and L. van der Torre. The pragmatic oddity in a norm-based semantics. In G. Governatori, editor, *16th International Conference in Artificial Intelligence and Law (ICAIL-17)*, pages 255–266. ACM, 217.

[27] H. Prakken and S. Marek. Contrary-to-duty obligations. *Studia Logica*, 57(1):91–115, 1996.

[28] A. K. Simpson. *Proof Theory and Semantics of Intuitionistic Modal Logic*. PhD thesis, University of Edinburgh, 1994.

[29] A. Stolpe. A theory of permission based on the notion of derogation. *Journal of Applied Logic*, 8(1):97–113, 2010.

[30] R. H. Thomason. On the strong semantical completeness of the intuitionistic predicate calculus. *Journal of Symbolic Logic*, 33(1):1–7, 1968.

[31] D. Wijesekera. Constructive modal logics I. *Annals of Pure and Applied Logic*, 50(3):271–301, 1990.

Received 21 October 2017

www.ingramcontent.com/pod-product-compliance
Lightning Source LLC
Chambersburg PA
CBHW081332090426
42737CB00017B/3110